D0082758

MODEL-BASED PREDICTIVE CONTROL

A Practical Approach

CONTROL SERIES

Robert H. Bishop
Series Editor
University of Texas at Austin
Austin, Texas

Published Titles

Linear Systems Properties: A Quick Reference
Venkatarama Krishnan

Robust Control Systems and Genetic Algorithms
*Mo Jamshidi, Renato A Krohling, Leandro dos Santos Coelho,
and Peter J. Fleming*

Sensitivity of Automatic Control Systems
Efim Rozenwasser and Rafael Yusupov

Model-Based Predictive Control: A Practical Approach
J.A. Rossiter

Forthcoming Titles

Material and Device Characterization Measurements
Lev I. Berger

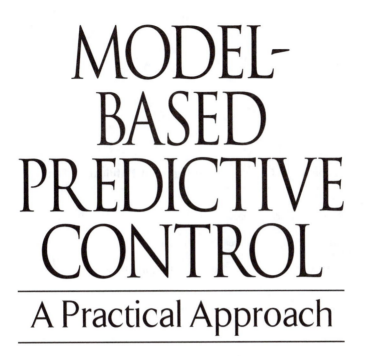

MODEL-BASED PREDICTIVE CONTROL

A Practical Approach

J. A. ROSSITER

CRC PRESS

Boca Raton   London   New York   Washington, D.C.

Library of Congress Cataloging-in-Publication Data

Rossiter, J.A.
    Model-based predictive control : a practical approach / by J.A. Rossiter.
        p. cm. — (Control series)
    Includes bibliographical references and index.
    ISBN 0-8493-1291-4 (alk. paper)
        1. Predictive control. 2. Control theory. I. Title. II. CRC Press control series

    TJ217.6.R67 2003
    629.8—dc21   2003048996

Visit the CRC Press Web site at www.crcpress.com

No claim to original U.S. Government works
International Standard Book Number 0-8493-1291-4
Library of Congress Card Number 2003048996
Printed in the United States of America 1 2 3 4 5 6 7 8 9 0
Printed on acid-free paper

Overview

The main aim of this book is to make the presentation less mathematically formal and hence more palatable for the less mathematically inclined. Insight is given in a non-theoretical way and there are a number of summary boxes to give a quick picture of the key results without the need to read through the detailed explanation.

The book can serve a twofold purpose: first as a textbook for graduate students and industrialists covering a detailed introduction to predictive control with a strong focus on the philosophy answering the questions, 'why?' and 'does it help me?' The basic concepts are introduced and then these are developed to fit different purposes: for instance, how to model, to give robustness, to handle constraints, to ensure feasibility, to guarantee stability and to consider what options there are with regard to models, algorithms, complexity versus performance, etc. The second role of the book is to target researchers in predictive control. In places the book goes into more depth, particularly in those areas where Dr. Rossiter has expertise.

In his research Dr. Rossiter has adopted a different style of presentation to that adopted by many authors and this style gives different insights to model-based predictive control. Dr. Rossiter uses this personal style and his own insight, hence forming a contrast to and complementing the other books available. Novel areas either not much discussed in other books or having recent developments are: (i) connections to optimal control and stability; (ii) the closed-loop paradigm; (iii) robust design in MPC; (iv) implementations of MPC using only small on-line computational burdens and (v) implicit modelling for predictive control.

Dr. Rossiter would like to apologise for any obvious references or topics that have been missed. He found writing a book a far more demanding task than anticipated and it was necessary to draw a line, at some point, on the continual improvement. Nevertheless, he does believe that this book complements the existing literature. By all means let him know of the large gaps you find and he will bear them in mind for a second edition.

Some MATLAB files have been written for readers of *Model-Based Predictive Control: A Practical Approach*. The files enable the user to design and simulate simple MPC controllers and moreover are easy to modify. They are available on the publisher's Web site at www.crcpress.com.

Acknowledgments

The main person I should thank is my friend and colleague, Prof. Kouvaritakis, who has had and continues to have a major influence on my work. Many of the insights in this book arose through working with Basil.

Other collaborators from whom I have learned much that is in this book are Mark Cannon, Jesse Gossner and Luigi Chisci. I would also like to thank Prof. Shah, who encouraged me to write a book, CRC Press for their patience while waiting for it, my family for their constant support and, of course, God for giving me the opportunity and skills to be who I am.

About the author

Dr. Rossiter has been researching predictive control since the late 1980s and he has published over 100 articles in journals and conferences on the topic. His particular contributions have focused on stability, feasibility and computational simplicity. He has a Bachelor's degree (1st class, 1987) and a doctorate (1990) from the University of Oxford. He spent 9 years as a lecturer at Loughborough University and is currently a reader at:

University of Sheffield
Department of Automatic Control and Systems Engineering
Mappin Street
Sheffield, S1 3JD
UK
email: J.A.Rossiter@shef.ac.uk

Contents

1

Introduction

This chapter gives a quick overview of the key concepts used in predictive control and introduces the notation that will be used in later chapters. The reader will find this useful to understand how predictive control works and why it can be so effective. Section 1.3 gives an overview of the structure of the book.

1.1 Overview of model-based predictive control

This book takes the view that we should concentrate our efforts on understanding the principles which unite most model-based predictive control (MPC) algorithms and not get distracted by the fine details which fill academic journals. Usually the details are self-evident given a clear understanding of the philosophy; these can always be obtained from the references. We will use the acronym MPC to denote all types of predictive control laws, for which many other abbreviations exist, e.g. IDCOM [97], DMC[23], GPC[21], QDMC[31], IMC[30], MUSMAR[39], RHC, to name just a few. The name is less important than the key characteristics.

Philosophically MPC reflects human behaviour whereby we select control actions which we think will lead to the best predicted outcome (or output) over some limited horizon. To make this selection we use an internal model of the process in question. We constantly update our decisions as new observations become available. Hence a predictive control law has the following components:

1. The control law depends on predicted behaviour.

2. The output predictions are computed using a process model.

3. The current input is determined by *optimising* some measure of predicted performance.

4. The receding horizon: the control input is updated at every sampling instant.

Summary: You will find MPC easier if at all times you try to understand the philosophy rather than the details. The details are obvious once the philosophy is understood. As your understanding of the details improves, read the introduction chapter again and ask yourself whether your insight is also improving.

1.2   The main components of MPC

This section discusses in more detail the main components of MPC and explains why these make sense. A good understanding of why we use these components can be used to gain insight into the action of different algorithms and hence how to modify a given algorithm to achieve specific aims.

1.2.1   Dependence of actions on predictions

Most control laws, say PID (proportional, integral and derivative), do not explicitly consider the future implication of current control actions. To some extent this is only accounted for by the expected closed-loop dynamics. MPC on the other hand explicitly computes the predicted behaviour over some horizon. One can therefore restrict the choice of the current *proposed* input trajectories to those that do not lead to difficulties in the future.

Think about crossing the road. It is not sufficient that the road has no cars on it between you and the other side; you also check whether there are cars still some distance away that will cross in front of you soon. That is, you predict whether at anytime during your crossing, you are likely to be hit by a car. If the answer is yes, you wait at the kerb. Moreover, all the time you are crossing, you keep looking, that is updating your predictions, so that if necessary your trajectory across the road can be altered.

Summary: Prediction is invaluable for avoiding otherwise unforeseen disaster.

1.2.2   Predictions are based on a model

In order to predict the future behaviour of a process, we must have a model of how the process behaves. In particular, this model must show the dependence of the output on the current measured variable and the current/future inputs. This model does not have to be linear (e.g. transfer function, state-space) and in fact can be just about anything.

As a human we often use fuzzy models and yet achieve very accurate control; for instance, if I am in third gear, doing 40 mph and not going up a steep incline, then depressing the accelerator should give good acceleration. The key point to note here is that a precise model is not always required to get tight control; because the decisions are updated regularly, this will deal with some model uncertainty in a fairly fast time scale. In the driving example above, failure to achieve good acceleration would be noted very quickly by comparison of actual behaviour with predicted model behaviour; for instance, am I in the correct gear, am I doing 40 and not 20, etc.? The decision on the best control is thus continually updated using information from this comparison.

In practice most MPC algorithms use linear models because the dependence of the predictions on future control choices is then linear and this faciliates optimisation as well as off-line analysis of expected closed-loop behaviour. However, nonlinear models can be used where the implied computational burden is not a problem and linear approximations are not accurate enough.

It is also important to note here the comment *fit for purpose*. In predictive control, the model is used solely to compute system output predictions, so the model is *fit for purpose* if it gives accurate enough predictions. The effort and detail put into modelling stage should reflect this. There may be no need to model all the physics, chemistry and internal behaviour of the process in order to get a model that gives reliable prediction, and in fact one should not model all this detail if it is not required. A basic rule base states that one should use the simplest model one can get away with.

> **Summary:** A model is used to generate system predictions. One should use the simplest model possible which is *fit for purpose*, that is, gives accurate enough predictions.

1.2.3 Selecting the current input

Before one can choose the current control action, one needs criteria to judge which action is best. Because MPC is usually implemented by computer, this requires a *numerical* definition so that a precise calculation can be made, that is, which predicted input trajectory gives the lowest numerical value to the *cost*. Selection of the *cost* function is an area of both engineering and theoretical judgement, but also gives rise to much debate in academia so we will not discuss that in the introduction.

Practical experience is that as long as some basic guidelines are followed, the actual choice of *cost* often has little effect on closed-loop performance. The main requirement is that the *cost* depends on the future controls and that low values of *cost* imply good closed-loop performance - good being defined for the process in question. Of course the choice of the *cost* affects the complexity of the implied optimisation and this is also a consideration*. For this reason 2-norm measures are popular, as the optimisation is straightforward.

---

*The extreme of this is predictive functional control (PFC); see Chapter 13.

Summary: The predicted inputs are selected as those minimising a given *cost* function. The *cost* function should be as simple as one can get away with for the desired performance.

1.2.4 Receding horizon

Consider again the process of driving. It is usually to consider the road several hundred yards ahead and to anticipate any potential dangers. As we move along the road, the part of the road within our vision moves with us so that we can always see the next few hundred yards. This in essence is the receding horizon, that is the limit of our vision (say 200 yds away) is always moving away at the same speed we are — it is receding. This means that we are continually picking up new information from the far horizon and this information is used to update our control actions (decisions). Predictive control works just like this: it considers the predicted behaviour over some horizon into the future and therefore at each successive sampling instant, it predicts one further sample into the future. As new information comes available the input trajectory is automatically modified to take account of it. The important question is however, how big should this horizon be? Using human intuition it is obvious that this horizon should be greater than the system settling time — otherwise one is ignoring behaviour with significant dynamics. Think about what would happen if when driving you only looked 20 yds ahead at 70 mph (i.e. well below braking distance) — you will surely crash or come off the road on the first sharp corner.

Summary: The horizon selected for predictions should include all significant dynamics (for instance, use the settling time); otherwise performance may be poor and important events may be unobserved.

1.2.5 Optimal or safe performance

In order to control a process very accurately, we need a very accurate model. Some insight into this can be given by racquet sports. When you are learning the sport, your mind has only an imprecise model of how arm movements, etc. affect the trajectory of the ball. As a result, control, direction, etc. are poor. As you practice, your internal model becomes more accurate and hence the quality of your play improves. This has significant repercussions on the complexity of future control strategies that can be considered. A novice uses simple strategies, as they have a simple model. Their goal is to keep the ball in play — they think only about the current move. An expert on the other hand may think several moves ahead and aim the ball very precisely in order to construct an opening with which to win the point. The same applies to MPC. There is no point selecting the next 10 control moves to optimise dynamic performance during transients if you have a poor model; you are better to go for a safe, perhaps slower option of ensuring that at least you move in the right direction and select only one move. Conversely if you do have an accurate model, then you could reliably select control moves over a larger future horizon and expect to get better dynamic performance; you would be foolish, sacrificing potential, to select only one move at a time.

Summary: You can only control as precisely as you can model! If you want a highly tuned controller, you need a very accurate model.

1.2.6 Tuning

There has been historical interest in the topic of tuning predictive control laws to ensure stability, much as there is much knowledge on how to tune PID. However, with MPC it is now understood that tuning is better allied to a specification of desired performance. If you get the *cost* function right, stability and tuning will look after themselves, as by definition you are optimising a *cost* which can only be small for good performance. When sensible guidelines are used, MPC will always give stable control (at least for the nominal) so the important judgements are how to get a balance between the performance in different loops, good sensitivity and also a balance between input activity and speed of response.

Classically these balances are achieved by weighting matrices, that is, putting different emphasis on performance in different loops according to their importance. It is doubtful however, that one could construct such weights systematically from a financial/operational viewpoint and in general on-line tuning is required until the balance looks right. A typical guideline to give a sensible initial value for the weights would be to normalise all the signals (input and output) so that the range 0 to 1 is equally important for each signal; then use unity weights on all the loops. It is difficult to generalise beyond this because each process has different priorities which may be too subtle to include in general guidelines.

Some authors [97] take the view that there should be no weighting of the inputs at all, as the goal is to drive the outputs to the desired point with whatever inputs are required. In this case it is paramount to set up the cost function carefully to avoid overtuned control laws and inversion of non-minimum phase characteristics. For instance, don't request unrealistically fast changes in the output and restrict the space of allowable input trajectories to those which move smoothly.

Summary: Tuning is often straightforward if one can define the relative importance of performance in different loops.

1.2.7 Constraint handling

One of the major selling points of MPC is its ability to do on-line constraint handling in a systematic way, hopefully retaining to some extent the stability margins and performance of the unconstrained law. The algorithm does this by optimising predicted performance subject to constraint satisfaction. For example, a racing driver optimises speed, subject to the constraint that the car remains on the track. If they optimised speed without taking explicit account of the constraint, their lap times would actually be far slower.

The details of how constraints are incorporated are very much dependent on the algorithm deployed and are discussed later. The algorithm selected will depend upon money available (i.e. related to potential increase in profit) and sampling time. The more you can spend, the closer to the true constrained optimum your algorithm is likely to get.

Summary: MPC takes systematic account of constraints and hence allows better performance.

1.2.8   Systematic use of future demands

When driving, if you see a corner ahead, you adjust the state (speed, gear, etc.) of the car so that you can take the corner safely. Typically a PID type of control law deals with the corner only once you reach it with the possible danger of going off the road (overshoot) due to excess entry speed. Feedforward can be added but the design could be somewhat ad hoc.

Conversely MPC automatically incorporates feedforward and moreover does this in a systematic way. The optimisation of the *cost* takes account of future changes in desired trajectory and measurable disturbances and so includes them as an intrinsic part of the overall control design[†].

Summary: MPC gives systematic feedforward design which is integrated with the constraint handling.

1.2.9   Systematic control design for multivariable systems

Another major selling point of MPC algorithms is that they can deal with multivariable (MIMO or multi-input-multi-output) systems in a systematic way. It is well known that PID (traditional control) design for highly interactive MIMO systems is very difficult. Although solutions have been developed through a combination of experience and time, these are often suboptimal and can be very detuned. Even tools like multivariable Nyquist [77, 79] often do not lead to a straightforward design.

The difficulty with tuning PID for MIMO systems actually relates back to the model; PID type designs make use of relatively little information about the plant and hence are unable to deal with the interaction effectively. MPC algorithms on the other hand demand and utilise a model (see Section 1.2.2) and thus make use of more informa-

---

[†]Be careful however; the default feedforward from MPC is not always a good one [109].

tion. This makes systematic design and analysis far easier. In MPC we specify a performance index so that in some sense one has a systematic design; that is, it gives the best performance by way of the *cost* specified for the model given. Experience has shown that this is an effective way to deal with interaction (MIMO problems) and moreover gives good bandwidth and gain/phase margins for the nominal case.

Summary: MPC allows systematic design for MIMO systems.

1.3 Overview of the book

The principal topics that are covered in this book are:

1. An overview of predictive control

2. Issues in prediction and identification for MPC

3. Stability analysis

4. Enhancing robustness

5. Constraint handling and feasibility issues

The idea is to build up an understanding of the different components discussed in Section 1.1 and to do this in a logical order. A brief summary of the chapters is given in the following list:

2 Introduces the model types used in this book. Models are a main building block of predictive control.

3 This considers how predictions are formed for a number of different models. Predictions are the second main building block of MPC.

4 This chapter introduces the performance index/optimisation which is the third main component of MPC and also shows how these are combined with a prediction model to form a control law.

5 Here the reader is shown a number of simple MPC designs and these are used to gain insight into systematic tuning rules and also scenarios to avoid.

6 The insights on tuning are developed further to show how one can set up an MPC law which is expected to give good performance. Moreover means of assuring stability, even when constraints are active, are given.

7 This chapter builds on the insights of the previous chapter and shows how to set up the MPC algorithm to be well conditioned. It introduces in detail the terminal set and overviews the options this brings.

8 Constraint handling is a major component of MPC. This chapter looks at potential problems with this and suggests some simple strategies.

9 This looks at the sensitivity of the unconstrained MPC law and introduces simple mechanisms for improving sensitivity.

10 This looks at the relationship between model structure and prediction errors and hence on loop sensitivity. It is noted that one can only control as accurately as one can predict.

11 This chapter looks at constraint handling in the uncertain case and introduces the tool of invariant sets.

12 At times one wants an MPC law with a small computational burden. Some insights into how to achieve this are given.

13 PFC is one example of a computationally simple MPC law that has achieved large acceptance in industry and hence is discussed here.

14 One advantage of MPC is its flexible structure. This chapter illustrates how this allows a systematic methodology for dealing with multirate systems.

15 A key component of MPC is the model. This chapter shows how the identification can be configured to support the MPC design.

A Gives some numerical examples that readers can use to test their own code and understanding. Also provides guidance on and examples of, typical questions that tutors may need.

For the basic user most of what they need will be in Chapters 1–6. A more advanced reader would be interested in Chapters 8–10 and perhaps 13. The remaining chapters introduce interesting areas and are more aimed at reseachers.

It is not the purpose of this book to write history but rather to state what is now understood and how to use this. Other books and many journal articles (e.g. [4, 13, 32, 84, 78, 144]) already give good historical accounts and so the reader is refered there for more details of how ideas developed. Also, it was necessary to draw a line somewhere so several important topics are not included, for instance: (i) MPC of continuous time models (sampled data systems); (ii) continuous time MPC; (iii) MPC of nonlinear systems. I suspect another book would be needed to do these three topics justice.

1.4 Notation

The following section is for reference so you are advised to come back to it as and when you need it. It is not intended for you to read this section first, as much of the notation will make more sense when you use it in the later chapters.

1.4.1 Table of notation

The most commonly used notation adopted in this book is given in Table 1.1. where as is normal bold lower case is used to denote vectors and non-bold lower case implies a scalar quantity. Capitals are used for matrices. Arguments such as $(.)(z)$ and $(.)(k)$ are often omitted to improve readability where their presence is implicit.

1.4.2 Vectors of past and future values

Given we need a notation for predictions, it is convenient to have a notation for a vector of future/past values of given variables. The notation of arrows pointing right is used for strictly future (not including current value) and arrows pointing left for past (including current value). The subscript denotes the sample taken as the base point, e.g.

$$\underset{\rightarrow k}{\mathbf{x}} = \begin{bmatrix} \mathbf{x}_{k+1} \\ \mathbf{x}_{k+2} \\ \vdots \end{bmatrix} ; \qquad \underset{\leftarrow k}{\mathbf{x}} = \begin{bmatrix} \mathbf{x}_k \\ \mathbf{x}_{k-1} \\ \vdots \end{bmatrix} \qquad (1.1)$$

The length of these vectors (that is, the time into the future or past) is not defined here and is usually implicit in the context.

1.4.3 Toeplitz and Hankel matrices

Toeplitz and Hankel matrices simplify much of the algebra commonplace in papers on MPC. They are simple to define. Consider a generic polynomial $n(z)$, where

$$n(z) = n_0 + n_1 z^{-1} + \cdots + n_m z^{-m} \qquad (1.2)$$

TABLE 1.1

Notation and variable names

Description	Typical notation	
Unit delay operator	z^{-1}	
Process outputs	y	**y**
Process inputs	u	**u**
Process incremental input ($u_k - u_{k-1}$)	Δu	Δ**u**
Process disturbance	d	**d**
Process states	**x**	
Process set point	r	**r**
Measurement noise	v	**v**
State-space matrices	A, B, C, D	
Delay operator (z-transforms)	z^{-1}	
Open-loop pole polynominal	$a(z)$ (or $d(z)$)	a
Closed-loop pole polynomial	$p_c(z)$	p_c
Open-loop zero polynomial	$b(z)$ (or $n(z)$)	b
Vector of coefficients of polynomial $n(z)$	**n**	
Controller numerator	$N_k(z)$	N_k
Controller denominator	$D_k(z)$	D_k
Difference operator	$\Delta(z) = 1 - z^{-1}$	Δ
State feedback gain	K	
Process model	$G(z)$	G
Toeplitz matrix of $n(z)$	C_n	Γ_n
Hankel matrix of $n(z)$	H_n	
Value of **x** at sampling instant k	$\mathbf{x}_k$	**x**(k)
Vector of future values of **x**	$\underset{\rightarrow}{\mathbf{x}}_k$	
Vector of past values of **x**	$\underset{\leftarrow}{\mathbf{x}}_k$	
Prediction matrices	H, P, Q	
Constraint matrices/vectors	C, E, **d**, **f**	
Upper and lower limts	$\overline{(.)}$, $\underline{(.)}$	

Then define the Toeplitz matrices Γ_n, C_n for $n(z)$ from the following stripped matrix

$$
\Gamma_n = \left[\frac{C_n}{M_n} \right] =
\begin{bmatrix}
n_0 & 0 & 0 & \cdots & 0 \\
n_1 & n_0 & 0 & \cdots & 0 \\
n_2 & n_1 & n_0 & \cdots & 0 \\
\vdots & \vdots & \vdots & \vdots & \vdots \\
n_m & n_{m-1} & n_{m-2} & & \vdots \\
\hline
0 & n_m & n_{m-1} & & \vdots \\
\vdots & \vdots & \vdots & \vdots & \vdots \\
0 & 0 & 0 & \cdots & n_0
\end{bmatrix}
\tag{1.3}
$$

Define the Hankel matrix H_n as

$$
H_n =
\begin{bmatrix}
n_1 & n_2 & n_3 & \cdots & n_{m-1} & n_m \\
n_2 & n_3 & n_4 & \cdots & n_m & 0 \\
n_3 & n_4 & n_5 & \cdots & 0 & 0 \\
\vdots & \vdots & \vdots & \vdots & \vdots & \vdots \\
n_{m-1} & n_m & 0 & \vdots & 0 & 0 \\
n_m & 0 & 0 & \vdots & 0 & 0 \\
0 & 0 & 0 & \vdots & 0 & 0 \\
\vdots & \vdots & \vdots & \vdots & \vdots & \vdots
\end{bmatrix}
\tag{1.4}
$$

In summary they have elements defined as

$$
\Gamma_n(i,j) = C_n(i,j) = n_{i-j}; \quad H_n(i,j) = n_{i+j-1}
\tag{1.5}
$$

where it is assumed that $n_i = 0$, $\begin{cases} i < 0 \\ i > m \end{cases}$.

NOTE: The dimensions of matrices Γ_n, C_n, H_n are not defined here as they are implicit in the context in which they are used. However, note the following:

1. Γ_n should have m more rows than columns but can have any number of columns.

2. C_n is always square but can be any dimension.

3. H_n has at least m non-zero rows and columns. You can add more zero rows and/or columns without changing its operation.

1.4.3.1 Polynomial multiplication

Γ_n is a tall and thin variant of the Toeplitz matrix C_n used for changing polynomial convolution into a matrix-vector multiplication. Define

$$
\begin{aligned}
d(z) &= d_0 + d_1 z^{-1} + \cdots + d_r z^{-r} \\
f(z) &= n(z)d(z) = f_0 + f_1 z^{-1} + \cdots + f_{n+r} z^{-n-r}
\end{aligned}
\tag{1.6}
$$

It is easy to show that the coefficients of $f(z)$ are given by

$$
\underbrace{\begin{bmatrix} f_0 \\ f_1 \\ f_2 \\ \vdots \\ f_m \\ f_{m+1} \\ \vdots \\ f_{n+r} \end{bmatrix}}_{\mathbf{f}} = \underbrace{\left[\begin{array}{ccccc} n_0 & 0 & 0 & \cdots & 0 \\ n_1 & n_0 & 0 & \cdots & 0 \\ n_2 & n_1 & n_0 & \cdots & 0 \\ \vdots & \vdots & \vdots & \vdots & \vdots \\ \hline n_m & n_{m-1} & n_{m-2} & \vdots & n_0 \\ 0 & n_m & n_{m-1} & \vdots & \\ \vdots & \vdots & \vdots & \vdots & \vdots \\ 0 & 0 & 0 & \cdots & n_m \end{array}\right]}_{\Gamma_n} \underbrace{\begin{bmatrix} d_0 \\ d_1 \\ d_2 \\ \vdots \\ d_r \end{bmatrix}}_{\mathbf{d}}
\tag{1.7}
$$

where in this case Γ_n has $r+1$ columns and $n+r+1$ rows. Alternatively one could write the answer as a polynomial, i.e.

$$
f(z) = \begin{bmatrix} 1 & z^{-1} & \cdots & z^{-n-m} \end{bmatrix}\mathbf{f} = \begin{bmatrix} 1 & z^{-1} & \cdots & z^{-n-m} \end{bmatrix}\Gamma_n\mathbf{d}
\tag{1.8}
$$

1.4.3.2 Inverting a Toeplitz matrix

One property of Toeplitz matrices is that inversion has a physical meaning. In summary

$$
[C_n]^{-1} = C_{\frac{1}{n}}
\tag{1.9}
$$

So for instance, if $\Delta = 1 - z^{-1}$ and $1/\Delta = 1 + z^{-1} + z^{-2} + \cdots$, then

$$
C_\Delta = \begin{bmatrix} 1 & 0 & 0 & \cdots & 0 \\ -1 & 1 & 0 & \cdots & 0 \\ 0 & -1 & 1 & \cdots & 0 \\ \vdots & \vdots & \vdots & \vdots & \vdots \\ 0 & 0 & 0 & \vdots & 1 \end{bmatrix} ; \quad [C_\Delta]^{-1} = C_{\frac{1}{\Delta}} = \begin{bmatrix} 1 & 0 & 0 & \cdots & 0 \\ 1 & 1 & 0 & \cdots & 0 \\ 1 & 1 & 1 & \cdots & 0 \\ \vdots & \vdots & \vdots & \vdots & \vdots \\ 1 & 1 & 1 & \vdots & 1 \end{bmatrix}
\tag{1.10}
$$

Hence one can form a relationship between quite complex matrix/vector multiplications and equivalent operations in transfer functions. For instance

$$
\mathbf{f} = [C_n]^{-1}\mathbf{d} \ \Rightarrow \ f(z) = \frac{d(z)}{n(z)}
\tag{1.11}
$$

where one assumes that the matrix dimensions are always taken to fit this context.

Summary: The relationship between matrix/vector multiplications and implied polynomial operations is very useful for analysis of predictions. Remember equations (1.8, 1.11).

1.4.3.3 Commutativity of Toeplitz matrices

If a multiplication is commutative, i.e. $f(z) = n(z)g(z)h(z) = h(z)g(z)n(z)$, then the corresponding Toeplitz operation is also commutative (with corresponding change of the dimensions of the Toeplitz matrices if required).

$$C_n C_g \mathbf{h} = C_g C_n \mathbf{h} = C_h C_g \mathbf{n}, \text{ etc.} \qquad (1.12)$$

Remark 1.1 *Although they have been defined with different dimensions, Γ_n, C_n are the same operator so we will tend to use C_n throughout in place of Γ_n to simplify the algebra.*

1.4.3.4 Toeplitz/Hankel matrices for multivariable systems

A further big advantage of using Hankel and Toeplitz matrices for prediction is that the multivariable case can be handled with no increase in algebraic complexity. Hence one can keep neat and compact algebra which aids insight and simplifies coding. A simple illustration is given here. Define two matrix polynomials (ones whose coefficients are matrices) as

$$
\begin{aligned}
N(z) &= N_0 + N_1 z^{-1} + N_2 z^{-2}; \\
D(z) &= D_0 + D_1 z^{-1} + D_2 z^{-2} + D_3 z^{-3}
\end{aligned}
\qquad (1.13)
$$

Then the corresponding Toeplitz and Hankel matrices for $N(z)$ are

$$
C_N = \begin{bmatrix}
N_0 & 0 & 0 & 0 & \cdots \\
N_1 & N_0 & 0 & 0 & \cdots \\
N_2 & N_1 & N_0 & 0 & \cdots \\
0 & N_2 & N_1 & N_0 & \cdots \\
\vdots & \vdots & \vdots & \vdots & \vdots
\end{bmatrix};
\quad
H_N = \begin{bmatrix}
N_1 & N_2 \\
N_2 & 0 \\
0 & 0 \\
\vdots & \vdots
\end{bmatrix}
\qquad (1.14)
$$

It is easy to show that the coefficients of the polynomial $F(z) = N(z)D(z)$ can be computed from

$$
F(z) = [I, Iz^{-1}, \ldots, Iz^{-5}]\mathbf{F}; \quad \mathbf{F} = C_N \mathbf{D}; \quad \mathbf{D} = \begin{bmatrix} D_0 \\ D_1 \\ D_2 \\ D_3 \end{bmatrix}
\qquad (1.15)
$$

Also one can show that (for square matrix polynomials)

$$[C_N]^{-1} = C_{N^{-1}} \qquad (1.16)$$

Remark 1.2 *Matrix multiplication is not commutative in general; that is, in general $N(z)D(z) \neq D(z)N(z)$.*

1.4.3.5 Rules for algebra of Hankel and Toeplitz matrices

We have been careful not to define explicitly the row and column dimensions of the Toeplitz and Hankel matrices. *This is deliberate and essential.* These matrices work like operators and hence the dimensions must be chosen appropriate to the context. For example, if one writes

$$\mathbf{f} = C_n \mathbf{g} \tag{1.17}$$

and this is intended to represent the operation $f(z) = n(z)g(z)$, then matrix C_n must have *at least as many* rows as the order of $n(z)g(z)$ plus one. Also its column dimension must match that of $\mathbf{g}$.

- One can give C_n too many rows without affecting the result. Too many rows will simply make $\mathbf{f}$ a longer vector, but the additional terms will all be zero, hence implying the same polynomial $f(z)$.

- One can give C_n too many columns as long as $\mathbf{g}$ is appended with zeros to give consistency of (1.17). Again additional zeros in $\mathbf{g}$ do not change the implied $g(z)$ and hence do not affect the implied operation.

- In practice when writing the algebra, one does not need to concern oneself with the precise implied dimensions.

Remark 1.3 *The same general observations will apply to Hankel matrices. If faced with a summation of the form*

$$H_f = H_n + H_g \tag{1.18}$$

then simply pack H_n or H_g with zeros to get dimensional compatibility. Zeros are always added at the bottom and/or on the right.

Summary:

- Treat C_n, H_n as operators and do not write their dimensions explicitly. Assume the implied minimal dimensions but allow them to be flexible to fit the context. The algebra of forming predictions is then far easier.

- There is a direct link between the matrix operations and z-transforms, e.g.

$$f(z) = g(z)h(z) = [1, z^{-1}, ..., z^{-r}]C_g \mathbf{h}; \quad \mathbf{g} = C_h^{-1}\mathbf{f} \; \Rightarrow \; g(z) = \frac{f(z)}{h(z)} \tag{1.19}$$

1.4.4 Common acronyms/abbreviations

This section lists most of the acronyms used in the book.

Abbreviation	Definition
ASM	Active set method
CARIMA	Controlled auto-regressive integrated moving average
CLP	Closed-loop paradigm
d.o.f.	degrees of freedom
DMC	Dynamic matrix control
DR	Dual rate
EMPC	Efficient model-based predictive control
EUM	Elimination of unstable modes
FIR	Finite impulse response
FR	Fast rate
GPC	Generalised predictive control
GPCI	Generalised predictive control with an independent model
GPCT	Generalised predictive control with a T-filter
I/O	Input/output
IC	Inferential control
IFT	Iterative feedback tuning
IHPC	Infinite horizon predictive control
IIR	Infinite impulse response
IMC	Internal model control
IM	Independent model
LMI	Linear matrix inequality
LP	Linear program
LQMPC	Linear quadratic optimal model predictive control
LTV	Linear time varying
MAS	Maximal admissible set
MFD	Matrix fraction description
MIMO	Multi-input-multi-output
MPC	Model predictive control
mph	miles per hour
MPQP	Multi parametric quadratic programming
MR	Multi-rate
NTC	No terminal control
OLP	Open-loop paradigm
ONEDOF	Algorithm using one d.o.f.
PFC	Predictive functional control

Abbreviation	Definition
PRBS	Pseudo random binary sequence
QP	Quadratic programming
RHC	Receding horizon control
SGPC	Stable generalised predictive control
SISO	Single-input-single-output
SR	Slow rate
s.t.	subject to
w.r.t.	with respect to
yds	yards

2

Common linear models used in model predictive control

In the first instance this book will make no apologies for giving just a summary statement about models. The art of modelling and the subtle differences between ARX, ARMA, and ARMAX models to name just a few is not a topic that is central to the theme of this book and is a subject covered extensively elsewhere (e.g. [76, 91]). However, the reader should be aware that the selection of the model is the most important part of an MPC design. Unexpectedly poor performance of an MPC controller will often be due to poor modelling assumptions. For this book (apart from Chapter 15) we will assume that the model is given and hence the purpose of this chapter is solely to show that different model types can be used in an MPC framework. How one can modify models to achieve benefits in an MPC design will be considered in later chapters.

You can use pretty much any model you like in an MPC strategy; however, if the model is nonlinear, then the implied optimisation may be nontrivial and moreover may not even converge. We will concentrate only on linear models and allow that any nonlinearity is mild and hence can be dealt with well enough by assuming model uncertainty and some gain scheduling of control laws. One book is not enough to cover the linear and the nonlinear case properly.

Summary:

1. For a detailed study of modelling the reader is referred elsewhere.

2. In MPC the aim is to choose the simplest model that gives accurate enough predictions.

2.1   Modelling uncertainty

2.1.1   Integral action and disturbance models

In this book the assumption will be made that there is a requirement for offset free tracking, as this forms the most common case and it would be difficult to generalise with the alternative assumption that offset is allowed. Hence it is also implicit that the control law must include integral action. Workers in MPC have developed convenient mechanisms for incorporating integral action and this chapter will illustrate this.

In the absence of uncertainty/disturbances one could easily obtain offset free control without integral action but in practice parameter uncertainty and disturbances necessitate the use of an integrator. In MPC the start point is different and one takes the viewpoint that:

Disturbance rejection is best achieved by having an internal model of the disturbance.

That is, if one models the disturbance appropriately, then the MPC control law can be set up to automatically reject the disturbance with zero offset. In fact the disturbance model most commonly used implicitly introduces an integrator into the control law and hence one also gets offset free tracking in the presence of parameter uncertainty. So the question to be answered in the modelling stage is, how are disturbance effects best included within the model?

The precise details of how to model disturbances is process dependent and if not straightforward is in the realm of a modelling specialist. However, that does not need to bother us here as long as we remember the following two simple guidelines.

1. In general we can control a process as accurately as we can model it. You need only improve your model if more accuracy is required.

2. There have been thousands of succesful applications of MPC using relatively simplistic assumptions for the disturbance model.

Summary: Assume that a simple disturbance model will be good enough. If the resulting control is unsatisfactory, revisit the modelling stage.

2.1.2   Modelling measurement noise

Similar statements can be made about modelling the noise (and other uncertainty in the process). It is usual to make simplistic assumptions rather than to define supposedly precise analytic answers. For instance, a Kalman filter will only give optimal observations if the model is exact and assumptions on the covariance of the uncertainty are also correct. In practice of course neither of these is true and one resorts to commonsense assumptions (process knowledge) and some on-line tuning.

In practice measurement noise is often assumed to be white and hence simply ignored. Coloured noise can be included in the models systematically but usually the corresponding filters (implicit in this) have other effects [158] and so are tuned with mixed objectives. Practical experience has shown that some form of low-pass filtering is nearly always required and/or beneficial but a systematic design of such filters for MPC is still an open question in general.

Summary: Filters (usually low-pass) are commonly needed to reduce sensitivity to measurement noise but it would be incorrect to say that typical designs reduce to an obvious optimal estimation problem.

2.2   Typical models

This section will give a brief summary of common linear models. The favoured model type depends very much on the reader and the process to be controlled and hence this book does not attempt to make a value judgement as to which is best. The focus is on the discrete time case, as MPC is usually implemented in discrete time.

Academia and the USA in particular, has put far more emphasis on state-space models. The advantage of these are that they extend easily to the multivariable case and there is a huge quantity of theoretical results which can be applied to produce controllers/observers and to analyse the models and resulting control laws.

Academics in Europe have also made extensive use of transfer function models and polynomial methods. Historically one advantage of this was the close relationship to popular black box identification techniques. However, this is much less an issue now with the development of subspace techniques for identifying black box state-space models [25]. The disadvantage of transfer function models is that their use in the multivariable case can be somewhat cumbersome and they are nonminimal representations. The advantage is that no state observer is required although one may argue the need to filter measurements implies the use of an observer in practice anyway.

Traditionally (this is changing now) industrialists have not favoured either of the above models in general and instead have used Finite impulse response (FIR) models e.g. [23, 97]. These are easy to understand and interpret, being for instance, the process step response. Although in practice these models could be determined by a single step test, the practical requirements for indentifying FIR models is that far more data is needed than to identify state-space and transfer function models and moreover there are issues of when to truncate the FIR and the need for unnecessarily large data storage requirements. FIR models however, do generally give lower sensitivity to measurement noise without the need for an observer (and associated design) and this can be a significant benefit for some industries.

MPC can also make use of the so-called independent model (IM) or internal model [30]. This can take the form of any model; the differences come in how it is used. This will be discussed later.

Summary: This book focuses on state-space models, transfer function models and FIR models.

2.3 State-space models

This section gives the terminology adopted in this book for representing state-space models and typical modelling assumptions used in MPC. The assumption will be made that the reader is familiar with state-space models.

2.3.1 Nominal state-space model

Using the notation of $(.)_k$ and $(.)(k)$ to imply a value at the kth sampling instant, the state-space model is given as:

$$
\underbrace{\begin{bmatrix} x_1(k+1) \\ x_2(k+1) \\ \vdots \\ x_n(k+1) \end{bmatrix}}_{\mathbf{x}_{k+1}} = \underbrace{\begin{bmatrix} a_{1,1} & a_{1,2} & \cdots & a_{1,n} \\ a_{2,1} & a_{2,2} & \cdots & a_{2,n} \\ \vdots & \vdots & \vdots & \vdots \\ a_{n,1} & a_{n,2} & \cdots & a_{n,n} \end{bmatrix}}_{A} \underbrace{\begin{bmatrix} x_1(k) \\ x_2(k) \\ \vdots \\ x_n(k) \end{bmatrix}}_{\mathbf{x}_k} + \underbrace{\begin{bmatrix} b_{1,1} & b_{1,2} & \cdots & b_{1,m} \\ b_{2,1} & b_{2,2} & \cdots & b_{2,m} \\ \vdots & \vdots & \vdots & \vdots \\ b_{n,1} & b_{n,2} & \cdots & b_{n,m} \end{bmatrix}}_{B} \underbrace{\begin{bmatrix} u_1(k) \\ u_2(k) \\ \vdots \\ u_m(k) \end{bmatrix}}_{\mathbf{u}_k}
$$

$$
\tag{2.1}
$$

$$
\underbrace{\begin{bmatrix} y_1(k) \\ y_2(k) \\ \vdots \\ y_l(k) \end{bmatrix}}_{\mathbf{y}_k} = \underbrace{\begin{bmatrix} c_{1,1} & c_{1,2} & \cdots & c_{1,n} \\ c_{2,1} & c_{2,2} & \cdots & c_{2,n} \\ \vdots & \vdots & \vdots & \vdots \\ c_{l,1} & c_{l,2} & \cdots & c_{l,n} \end{bmatrix}}_{C} \underbrace{\begin{bmatrix} x_1(k) \\ x_2(k) \\ \vdots \\ x_n(k) \end{bmatrix}}_{\mathbf{x}_k} + \underbrace{\begin{bmatrix} d_{1,1} & d_{1,2} & \cdots & d_{1,m} \\ d_{2,1} & d_{2,2} & \cdots & d_{2,m} \\ \vdots & \vdots & \vdots & \vdots \\ d_{l,1} & d_{l,2} & \cdots & d_{l,m} \end{bmatrix}}_{D} \underbrace{\begin{bmatrix} u_1(k) \\ u_2(k) \\ \vdots \\ u_m(k) \end{bmatrix}}_{\mathbf{u}_k}
$$

In abbreviated form the model is

$$
\mathbf{x}_{k+1} = A\mathbf{x}_k + B\mathbf{u}_k; \quad \mathbf{y}_k = C\mathbf{x}_k + D\mathbf{u}_k \tag{2.2}
$$

$\mathbf{x}$ denotes the state vector (dimension n), $\mathbf{y}$ (dimension l) denotes the process outputs (or measurements) to be controlled and $\mathbf{u}$ (dimension m) denotes the process inputs (or controller output) and A, B, C, D are the matrices defining the state-space model. Ordinarily for real processes $D = 0$.

2.3.2 Nonsquare systems

Although MPC can cope with nonsquare systems ($l \neq m$), it is more usual to do some squaring down and hence control a square system. When MPC is applied directly to a nonsquare system the precise objectives and associated tuning are process dependent and nongeneric; hence we omit this topic. In simple terms if $m > l$, several inputs can be used to achieve the same output and so the optimisation must be set up to make optimal use of the spare degrees of freedom (d.o.f.), the definition of optimum being process dependent. If $l > m$, there are too few d.o.f. and so we must accept offset in some output loops; in this case additional criteria are required to set up the control strategy and again these are process dependent.

2.3.3 Including a disturbance model

A good discussion of this can be found in [90]. First decide whether the disturbance is a simple perturbation to the output or affects the states directly. We will treat each in turn.

2.3.3.1 Output disturbance

A common model of the disturbance is given as integrated white noise. That is, disturbance $\mathbf{d}_k$ is modelled as

$$\mathbf{d}_{k+1} = \mathbf{d}_k + v_k \tag{2.3}$$

where v_k is unknown and zero mean. It is assumed throughout that $\mathbf{d}_k$ is also unknown though of course it can be partly inferred via an observer.

Such a disturbance can be incorporated into the state-space model by replacing (2.2b) as follows:

$$\mathbf{y}_k = C\mathbf{x}_k + D\mathbf{u}_k + \mathbf{d}_k \tag{2.4}$$

Of course the disturbance is unknown, as is the state $\mathbf{x}_k$, so it must be estimated. By including the disturbance into the system dynamics, the assumed process model becomes

$$\mathbf{z}_{k+1} = \tilde{A}\mathbf{z}_k + \tilde{B}\mathbf{u}_k; \quad \mathbf{y}_k = \tilde{C}\mathbf{z}_k + D\mathbf{u}_k + v_k \tag{2.5}$$

where

$$\mathbf{z}_{k+1} = \begin{bmatrix} \mathbf{x}_{k+1} \\ \mathbf{d}_{k+1} \end{bmatrix}; \quad \tilde{A} = \begin{bmatrix} A & 0 \\ 0 & I \end{bmatrix}; \quad \tilde{B} = \begin{bmatrix} B \\ 0 \end{bmatrix}; \quad \tilde{C} = \begin{bmatrix} C & I \end{bmatrix} \tag{2.6}$$

An observer can be constructed for this model, under the usual assumption of observability, to give estimates of both the state $\mathbf{x}$ and disturbance $\mathbf{d}$.

2.3.3.2 State disturbance

In this case one still uses assumption (2.3) but it is included in the state update equation; that is, replace (2.2) by

$$\mathbf{x}_{k+1} = A\mathbf{x}_k + B\mathbf{u}_k + F\mathbf{d}_k; \quad \mathbf{d}_{k+1} = \mathbf{d}_k + v_k; \quad \mathbf{y}_k = C\mathbf{x}_k + D\mathbf{u}_k \qquad (2.7)$$

Again, the overall process model should be augmented to include the disturbance dynamic as follows:

$$\mathbf{z}_{k+1} = \tilde{A}\mathbf{z}_k + \tilde{B}\mathbf{u}_k; \quad \mathbf{y}_k = \tilde{C}\mathbf{z}_k + D\mathbf{u}_k + v_k \qquad (2.8)$$

where

$$\mathbf{z}_{k+1} = \begin{bmatrix} \mathbf{x}_{k+1} \\ \mathbf{d}_{k+1} \end{bmatrix}; \quad \tilde{A} = \begin{bmatrix} A & F \\ 0 & I \end{bmatrix}; \quad \tilde{B} = \begin{bmatrix} B \\ 0 \end{bmatrix}; \quad \tilde{C} = \begin{bmatrix} C & 0 \end{bmatrix} \qquad (2.9)$$

Summary: The state-space model of a process incorporating a disturbance model can be given by either (2.5) or (2.8) as appropriate.

2.3.4 Systematic inclusion of integral action with state-space models

Typical state feedback does not incorporate integral action. This section shows one method by which this limitation can be overcome.

2.3.4.1 Offset with typical state feedback

Let us assume hereafter that for all real processes there is no instantaneous feed through from the input to the output, that is $D = 0$. Then, even with a zero setpoint, the typical stabilising state feedback of the form

$$\mathbf{u}_k = -K\mathbf{x}_k \qquad (2.10)$$

will not give offset free control in the presence of nonzero disturbances. This is self evident from substitution of (2.10) into, for example (2.2, 2.4), which implies

$$\lim_{k\to\infty} \mathbf{x}_k = 0 \quad \Rightarrow \quad \lim_{k\to\infty} \mathbf{y}_k = \mathbf{d}_k \qquad (2.11)$$

2.3.4.2 A form of state feedback giving no offset

Consider now [90] an alternative form of state feedback

$$\mathbf{u} = -K(\mathbf{x} - \mathbf{x}_{ss}) + \mathbf{u}_{ss} \qquad (2.12)$$

where $\mathbf{x}_{ss}$, $\mathbf{u}_{ss}$ are estimates of the steady-state values of state and input giving offset free tracking. Under the assumption that $\mathbf{x}_{ss}$, $\mathbf{u}_{ss}$ are consistent, then for fixed $\mathbf{d}_k$, such a control law will necessarily drive

$$\lim_{k\to\infty} \begin{cases} \mathbf{x}_k = \mathbf{x}_{ss} \\ \mathbf{u}_k = \mathbf{u}_{ss} \end{cases}$$

Again, this is self-evident by substitution of (2.12) into (2.2, 2.4) or (2.7). Clearly offset free tracking follows automatically if

$$\left.\begin{array}{c} \mathbf{x}_k = \mathbf{x}_{ss} \\ \mathbf{u}_k = \mathbf{u}_{ss} \end{array}\right\} \quad \Rightarrow \quad \mathbf{y}_k = \mathbf{r} \tag{2.13}$$

where $\mathbf{r}$ is the set point.

Summary: Incorporating integral action into a state feedback is equivalent to finding consistent estimates of the steady-state values of the state and input; that is, $\mathbf{x}_{ss}, \mathbf{u}_{ss}$.

2.3.4.3 Estimating steady-state values of the state and input

In order to implement control law (2.12), we need a means of estimating mutually consistent values for the observer states $\mathbf{x}_{ss}, \mathbf{u}_{ss}$. First assume observability so that state estimates converge and are mutually consistent by way of the model (even if due to model uncertainty they do not match the true process exactly). Define the desired output as set point $\mathbf{r}$.

1. Assume (see 2.4) that the current estimate of $\mathbf{d}$ is the best estimate of its value in the future (i.e. $E[v_k] = 0$).

2. Given that $\mathbf{d}$, $\mathbf{r}$ are known, one can estimate the required steady-state values $(\mathbf{x}_{ss}, \mathbf{u}_{ss})$ of $\mathbf{x}, \mathbf{u}$ to get the correct output from the relevant consistency conditions (equations (2.2, 2.4) or (2.7) respectively, for output and state disturbances).

$$\left\{\begin{array}{l} \mathbf{r} = C\mathbf{x}_{ss} + \mathbf{d} \\ \mathbf{x}_{ss} = A\mathbf{x}_{ss} + B\mathbf{u}_{ss} \end{array}\right\} \quad \text{or} \quad \left\{\begin{array}{l} \mathbf{r} = C\mathbf{x}_{ss} \\ \mathbf{x}_{ss} = A\mathbf{x}_{ss} + B\mathbf{u}_{ss} + F\mathbf{d} \end{array}\right\} \tag{2.14}$$

3. These are simple simultaneous equations and give a solution of the form

$$\begin{bmatrix} \mathbf{x}_{ss} \\ \mathbf{u}_{ss} \end{bmatrix} = M \begin{bmatrix} \mathbf{r} \\ \mathbf{d} \end{bmatrix} \tag{2.15}$$

where matrix M depends solely on matrices A, B, C, F representing the model.

Remark 2.1 *One can also set up the disturbance model on the states or inputs ($F \neq 0$) if that is more appropriate; details are in [90]. Moreover it is easy to show that the above integral action is robust to model uncertainty, as it is based solely on consistency of the observer equations (i.e. the model), which are known exactly.*

Summary: Substitution of (2.15) into (2.12) ensures integral action within a state feedback control law.

2.4   Transfer function models (single-input/single-ouput)

A popular model [21] is the so-called Controlled auto-regressive integrated moving average (CARIMA) model. This can capture most of the variability in transfer function models by a suitable selection of parameters. It is given as

$$a(z)y_k = b(z)u_k + \frac{T(z)}{\Delta(z)}v_k \qquad (2.16)$$

where v_k is an unknown zero mean random variable which can represent disturbance effects and measurement noise simultaneously. Although there exist modelling techniques to compute *best fit* values for the parameters of $a(z), b(z)$ and $T(z)$, in MPC it is commonplace to treat $T(z)$ as a design parameter (e.g. [21, 158]); this because it has direct effects on loop sensitivity and so one may get better closed-loop performance with a $T(z)$ which is notionally not the best fit. It is common to write transfer function models in the equivalent difference equation form. For instance with $T = 1$, (2.16) is given as

$$y_{k+1} + a_1 y_k + \cdots + a_n y_{k-n+1} = b_1 u_k + b_2 u_{k-1} + \cdots + b_n u_{k-n+1} + d_{k+1} \qquad (2.17)$$

where $a(z) = 1 + a_1 z^{-1} + \cdots + a_n z^{-n}$, $b(z) = b_1 z^{-1} + \cdots + b_n z^{-n}$ and d_k is the unknown disturbance term derived from $d_k = \dfrac{T(z)}{\Delta(z)} v_k$.

2.4.1   Disturbance modelling

Recall the earlier statement *disturbance rejection is best achieved by having an internal model of the disturbance.* Then notice that the choice of $T(z) = 1$ gives an equivalence to (2.3):

$$d_k = \frac{1}{\Delta(z)} v_k \quad \equiv \quad d_{k+1} = d_k + v_k \qquad (2.18)$$

Hence it is clear that the term $(T(z)/\Delta(z))v_k$ deployed in (2.16) is a disturbance model and is similar to that implied in (2.4, 2.7). The term $d = v/\Delta$ represents an integrated white noise term or a random walk; this is a well accepted model for disturbances, as it allows nonzero mean with random changes. The choice of T determines equivalence to either (2.4) or (2.7) or other possibilities.

Summary: The CARIMA model allows systematic inclusion of a disturbance models which therefore facilitates affective disturbance rejection.

2.4.2   Consistent steady-state estimates with CARIMA models

As seen in Section 2.3, the key to achieving integral action in an MPC control law is a consistent and correct assessment of the expected steady-state value of the input (the state is not used for transfer function models) such that one gets offset free tracking. Clearly the desired output is the set point, but the corresponding value of the input depends upon the unknown disturbance. Obviously, assuming that $d_k = d_{k-1}, v_k = 0$, then d_k can be inferred by writing (2.17) at the current and previous sampling instants and solving from the two implied simultaneous equations.

However in MPC it is usual to use a different method. Write an incremental version of (2.16); that is, relate the outputs to control increments $\Delta u_k = u_k - u_{k-1}$. This is equivalent to either: (i) multiplying (2.16) by Δ or (ii) subtracting (2.17) at $k-1$ from (2.17) at k. Hence

$$[a(z)\Delta(z)]y(z) = b(z)\Delta u(z) + T(z)v \tag{2.19}$$

Clearly this operation has eliminated the nonzero mean unknown variable d_k and the only remaining unknown, v_k, is zero mean, can be assumed zero in the future, and hence does not affect predictions.

A second and equally useful benefit of using (2.19) instead of (2.17) is that the input is now written in terms of increments and clearly in steady-state the increments will be zero. That is

$$\lim_{k \to \infty} \Delta u_k = \Delta u_{ss} = 0 \tag{2.20}$$

Hence the consistent estimates of the states required to give offset free tracking in the steady-state are $y = r$, $\Delta u = 0$.

> **Summary:** One can get consistent steady-state predictions with the incremental model (2.19). It is convenient to define $A(z) = a(z)\Delta(z)$.

2.4.3   Achieving integral action with CARIMA models

The details of this will be more transparent after later sections. One needs to assume the form of the control law (much as we assumed (2.12)) before we can establish how to ensure integral action. It is known that MPC control laws based on transfer function models depend upon the predictions and hence must take the form:

$$D_k(z)\Delta u = P_r(z)r - N_k(z)y \tag{2.21}$$

It is clear from the presence of the $D_k\Delta$ term that there is an integrator in the forward path and hence disturbances will be rejected. Furthermore in order to get no tracking offset in the steady state one must check consistency of the following steady-state conditions:

$$\{\Delta u_k = 0; \quad y_k = r\} \ \forall k \tag{2.22}$$

Clearly this implies

$$P_r(1) = N_k(1) \tag{2.23}$$

Summary: The control law (2.21) has integral action and gives no offset to non-zero set points if $P_r(1) = N_k(1)$.

2.4.4 Selection of $T(z)$ for MPC

Although T is a design polynominal used to *model* the disturbance signal and hence improve disturbance rejection, it can also be used to enhance robustness of the closed-loop. However, the design of T is not systematic in general and a few basic guidelines only are given (e.g. see [158]):

1. Let $T = \hat{a}\hat{T}$, where $\hat{a}$ contains the dominant system pole/poles and $\hat{T}$ contains further poles near one. For a system sampled at typical rates $T = [1 - 0.8z^{-1}]^n$ works fairly well.

2. Design $1/T$ to be a low-pass filter which removes high frequency noise but not the dominant frequencies in the model.

3. Use some trial and error. If $T = 1$ works well, then you may get little benefit from more complicated T.

4. The choice of $T = a$ has equivalence to the use of an FIR model.

5. Systematic designs for T are nonlinear and therefore not simple.

Summary: More discussion on $T(z)$ is found in Chapter 9 on loop robustness.

2.5 FIR models

Historically these were the most common model form encountered in industrial MPC packages although that is beginning to change.

2.5.1 Impulse response models

Take the model with inputs, outputs and disturbance $\mathbf{u}, \mathbf{y}, \mathbf{d}$, respectively,

$$\mathbf{y}_k = G(z)\mathbf{u}_k + \mathbf{d}_k \tag{2.24}$$

Then the process $G(z)$ (for the stable case which predominates in practice) can be represented by a Tayler series in the delay operator, that is

$$G(z) = \sum_{i=0}^{\infty} G_i z^{-1} \tag{2.25}$$

Equivalently this sequence can be viewed as the impulse response, the expected values of the output **y** in the event **u** comprises a single impulse. As with transfer function models, the parameters G_i can be identified using standard methods.

Disadvantages: Impulses are not used in practice so the sequence does not lend itself to intuition. Also there may be a need for a very large number of terms for adequate convergence.

Advantages: One can avoid issues like selection of the model order and identification of the deadtime.

2.5.2   Step response models

A step response model is a sequence of values representing the step response of the process and can be written as

$$H(z) = \sum_{i=0}^{\infty} H_i z^{-i} \tag{2.26}$$

where clearly $H(z) = G(z)/\Delta(z)$. The corresponding input/output equation can be derived from (2.24) and takes the form

$$\mathbf{y}_k = H(z)\Delta\mathbf{u}_k + \mathbf{d}_k \tag{2.27}$$

The logic for getting offset free prediction follows the same lines as that given in the previous section; that is, subtract (2.27) at the previous sample from the current to eliminate the unknown $\mathbf{d}_k$. This will be discussed in more detail in the following chapter.

Disadvantages: Include needing a very large number of terms for adequate convergence. Moreover $\lim_{i\to\infty} H_i \neq 0$ in general.

Advantages: Avoiding issues like selection of model order and identification of deadtime and being intuitive/easy to understand.

2.6   Independent models

Independent models (IM) are not a different form of model; however, it is important to include a short discussion in this chapter because there is a key difference in

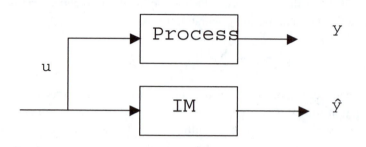

FIGURE 2.1
Independent model.

the philosophy of how the model is used. The difference in usage is particularly relevant in the context of MPC, as it can give substantial changes in the closed-loop sensitivity, although not in nominal performance. Hence it is an alternative that should be considered. It has connections with Smith predictor ideas and internal model control (IMC) [30].

- The IM approach is not restricted to any given model type. Use whichever is easiest.

- Its use is equivalent to an FIR model without truncation errors.

- Even if IM is a state-space model, an observer is not required!

This topic will be considered more carefully in the next chapter on prediction, as MPC uses the model solely to form predictions. For now it is sufficient to note that the IM is a process model which is simulated in parallel with the process using the same inputs; see Figure 2.1. If the process has output y, then the IM has process $\hat{y}$ which in general will be different, but similar.

2.7   Matrix-fraction descriptions

Although state-space models are usually favoured for the multi-input/multi-output (MIMO) case, an observer is still required which is not the case with transfer function models (although some may argue any filtering is equivalent). A MIMO transfer function model can be represented as a matrix fraction description (MFD) e.g.

$$D(z)\mathbf{y}(z) = N(z)\mathbf{u}(z) + \frac{T(z)\mathbf{v}}{\Delta} \tag{2.28}$$

where $N(z)$, $D(z)$ are matrix polynomials in the delay operator. In difference equation form (with $T = 1$) this gives

$$\begin{aligned}
\mathbf{y}_{k+1} &= N_1\Delta\mathbf{u}_k + N_2\Delta\mathbf{u}_{k-1} + \cdots + N_l\Delta\mathbf{u}_{k-l} \\
&\quad -D_1\mathbf{y}_k - D_2\mathbf{y}_{k-1} - \cdots - D_m\mathbf{y}_{k-m}
\end{aligned} \tag{2.29}$$

Hereafter in the context of MPC, these models can be used in the same way as SISO CARIMA models, so long as one remembers that matrix multiplication is not commutative in general. FIR models are particular forms of MFD with $D(z) = I$.

Disadvantages: Selection of model order/dead time to give a numerically robust identification may not be straightforward. It is likely to contain implied near pole/zero cancellations. Minimum order model is not unique.

Advantages: It is easy to identify with standard packages and relatively small number of parameters. No observer is required.

Remark 2.2 *Recent work on subspace identification (e.g. [27]) and multi-models [126] has strong links with MFD models.*

2.8 Modelling the dead times in a process

In work that concentrates solely on modelling as an end in itself, a knowledge of or ability to infer, the system dead time is important. It is well known from simple Nyquist stability analysis that dead times make systems harder to control and moreover a mismatch between the assumed and actual dead time can make large differences to control quality. A classical solution for controlling systems with large dead times is to deploy a Smith predictor to simulate the expected value of the process in the future and control an offset corrected version of this rather than the plant. This gives equivalent stability margins to the process without a dead time but the approach is of course sensitive to errors in the dead time.

MPC uses a similar strategy in that by predicting the behaviour at future points it can deal with process dead time systematically. However, there is a key difference from the Smith predictor which improves robustness of MPC and implies that an exact estimate of the dead time is not as critical. The control law calculation is based on a whole trajectory, not just a single point, hence the emphasis is placed more on where the responses settle rather than transients. The implication for modelling is that in the identification stage one can afford to *underestimate* the dead time and let the identification algorithm insert small values against coefficients that perhaps should be zero (as long as one avoids near cancellation of unstable pole/zero pairs). This simplifies modelling and has a negligible effect on control design.

For instance, if the true plant were

$$G = \frac{z^{-3}(1 - 2z^{-1})}{1 - 0.8z^{-1}} \qquad (2.30)$$

then using

$$G = \frac{\varepsilon z^{-2} + z^{-3}(1 - 2z^{-1})}{1 - 0.8z^{-1}} \qquad (2.31)$$

even where ε is not small would often be fine in practice.

Summary: In general it is quite safe to underestimate dead time when modelling for an MPC control law, assuming of course reasonable estimates of model order.

3

Prediction in model predictive control

MPC algorithms make use of predictions of the system behaviour. In this chapter it will be shown how to compute those predictions for various types of linear models. For the reader who is not interested in the detail, simply read the results which will be highlighted at the end of each section.

> **Main observation:** For linear models, the predictions of future outputs are affine in the current state and the future control increments.

3.1 General format of prediction modelling

Many papers concentrate on the details to such an extent that the *simplicity* of the main result is lost. This section is intended to draw attention to it.

Define the systems state as **x** (this could consist of past input/output data); then a general form of future predictions is

$$\underset{\rightarrow k}{\mathbf{y}} = H\underset{\rightarrow k-1}{\Delta\mathbf{u}} + P\underset{\leftarrow k}{\mathbf{x}} \qquad (3.1)$$

- H is the Toeplitz matrix $C_{G/\Delta}$ (see Section 1.4.3) of the system step response.

- P is a matrix whose coefficients depend on the model parameters in a straight-forward but nonlinear manner.

- Look at Section 1.4 to remind yourself of the arrow notation for vectors of future/past values.

For many users the details of how to compute H, P are not important and it is suffi-cient to accept that this is trivial with linear models.

> **Summary:** Once you are happy that equation (3.1) is true, you can proceed to the next chapter. Continue with this chapter to find out more details.

3.2 Prediction with state-space models

Prediction with a state-space model is straightforward. The details are given next. Consider the state-space model which gives the one-step ahead predictions:

$$\mathbf{x}_{k+1} = A\mathbf{x}_k + B\mathbf{u}_k; \quad \mathbf{y}_{k+1} = C\mathbf{x}_{k+1} \tag{3.2}$$

One can use this relationship recursively to find predictions, for instance:

1. Write (3.2) at $k+2$

$$\mathbf{x}_{k+2} = A\mathbf{x}_{k+1} + B\mathbf{u}_{k+1}; \quad \mathbf{y}_{k+2} = C\mathbf{x}_{k+2} \tag{3.3}$$

2. Subsitute (3.2) into (3.3) to eliminate $\mathbf{x}_{k+1}$

$$\mathbf{x}_{k+2} = A^2\mathbf{x}_k + AB\mathbf{u}_k + B\mathbf{u}_{k+1}; \quad \mathbf{y}_{k+2} = C\mathbf{x}_{k+2} \tag{3.4}$$

3. Next write the prediction at $k+3$ using the two-step ahead prediction (3.4), i.e.

$$\mathbf{x}_{k+3} = A^2\mathbf{x}_{k+1} + AB\mathbf{u}_{k+1} + B\mathbf{u}_{k+2}; \quad \mathbf{y}_{k+3} = C\mathbf{x}_{k+3} \tag{3.5}$$

and again substitute from (3.2) to eliminate $\mathbf{x}_{k+1}$:

$$\mathbf{x}_{k+3} = A^2[A\mathbf{x}_k + B\mathbf{u}_k] + AB\mathbf{u}_{k+1} + B\mathbf{u}_{k+2}; \quad \mathbf{y}_{k+3} = C\mathbf{x}_{k+3} \tag{3.6}$$

4. More generally one can continue this recursion to give the n-step ahead predictions as:

$$
\begin{aligned}
\mathbf{x}_{k+n} &= A^n\mathbf{x}_k + A^{n-1}B\mathbf{u}_k + A^{n-2}B\mathbf{u}_{k+1} + \cdots + B\mathbf{u}_{k+n-1} \\
\mathbf{y}_{k+n} &= C[A^n\mathbf{x}_k + A^{n-1}B\mathbf{u}_k + A^{n-2}B\mathbf{u}_{k+1} + \cdots + B\mathbf{u}_{k+n-1}]
\end{aligned} \tag{3.7}
$$

Hence one can form the whole vector of future predictions up to a horizon n_y as follows:

$$
\underbrace{\begin{bmatrix} \mathbf{x}_{k+1} \\ \mathbf{x}_{k+2} \\ \mathbf{x}_{k+3} \\ \vdots \\ \mathbf{x}_{k+n_y} \end{bmatrix}}_{\underset{\rightarrow k}{\mathbf{x}}} = \underbrace{\begin{bmatrix} A \\ A^2 \\ A^3 \\ \vdots \\ A^{n_y} \end{bmatrix}}_{P_{xx}} \mathbf{x}_k + \underbrace{\begin{bmatrix} B & 0 & 0 & \cdots \\ AB & B & 0 & \cdots \\ A^2B & AB & B & \cdots \\ \vdots & \vdots & \vdots & \vdots \\ A^{n_y-1}B & A^{n_y-2}B & A^{n_y-3}B & \cdots \end{bmatrix}}_{H_x} \underbrace{\begin{bmatrix} \mathbf{u}_k \\ \mathbf{u}_{k+1} \\ \mathbf{u}_{k+2} \\ \vdots \\ \mathbf{u}_{k+n_y-1} \end{bmatrix}}_{\underset{\rightarrow k-1}{\mathbf{u}}} \tag{3.8}
$$

and

$$
\underbrace{\begin{bmatrix} \mathbf{y}_{k+1} \\ \mathbf{y}_{k+2} \\ \mathbf{y}_{k+3} \\ \vdots \\ \mathbf{y}_{k+n_y} \end{bmatrix}}_{\underset{\rightarrow k}{\mathbf{y}}} = \underbrace{\begin{bmatrix} CA \\ CA^2 \\ CA^3 \\ \vdots \\ CA^{n_y} \end{bmatrix}}_{P} \mathbf{x}_k + \underbrace{\begin{bmatrix} CB & 0 & 0 & \ldots \\ CAB & CB & 0 & \ldots \\ CA^2B & CAB & CB & \ldots \\ \vdots & \vdots & \vdots & \vdots \\ CA^{n_y-1}B & CA^{n_y-2}B & CA^{n_y-3}B & \ldots \end{bmatrix}}_{H} \underset{\rightarrow k-1}{\mathbf{u}} \tag{3.9}
$$

Remark 3.1 *These prediction equations do not consider the disturbance model explicitly. However, it can be included in a manner discussed in Section 2.3.4. That is, use predictions (3.10) to find the nominal state feedback and then adapt the feedback as in (2.12) to ensure offset free control.*

Summary: The predictions for model (3.2) are

$$
\begin{aligned}
\underset{\rightarrow k}{\mathbf{x}} &= P_{xx}\mathbf{x}_k + H_x \underset{\rightarrow k-1}{\mathbf{u}} \\
\underset{\rightarrow k}{\mathbf{y}} &= P\mathbf{x}_k + H \underset{\rightarrow k-1}{\mathbf{u}}
\end{aligned} \tag{3.10}
$$

A worked example is given in Section A.1.1.

3.3 Prediction with transfer function models – matrix methods

The first point to note here is that there are several different ways of deriving the prediction equations for transfer function models (e.g. [102]); some are more transparent than others. This book will not contrast the different methods in detail because they give the same prediction equations so which to use reduces to personal preference.

Typically papers in the academic journals make use of diophantine identities to form the prediction equations. However, this procedure tends to obscure what is actually going on, that is prediction, and hence can be confusing for the newcomer. Moreover the historical reason for such a preference is unclear. Instead here, use will be made of matrix methods which are much easier to relate to prediction and hence easier to understand and code. The cases of $T(z) = 1$ and $T(z) \neq 1$ will be tackled in turn.

3.3.1 Prediction for a CARIMA model with $T(z) = 1$ – SISO case

As discussed in Section 2.4.2, the CARIMA model is first replaced by its incremental form, as this allows offset free prediction in the steady state and also one can ignore the unknown term. Hence let the original nominal model be given as

$$a(z)y_k = b(z)u_k \tag{3.11}$$

Then, with $A(z) = a(z)\Delta(z)$, the incremental form of this equation is

$$A(z)y_k = b(z)\Delta u_k \tag{3.12}$$

Define $A(z), b(z)$ as follows where it is assumed that for a strictly proper system $b_0 = 0$.

$$\begin{aligned}
A(z) &= 1 + A_1 z^{-1} + A_2 z^{-2} + \cdots + A_{n+1} z^{-n-1} \\
b(z) &= b_1 z^{-1} + b_2 z^{-2} + \cdots + b_n z^{-n}
\end{aligned} \tag{3.13}$$

3.3.1.1   One-step ahead prediction with $T(z) = 1$ – SISO case

The corresponding difference equation (or one step ahead prediction) for model (3.12) is:

$$y_{k+1} = -A_1 y_k - \cdots - A_{n+1} y_{k-n} + b_1 \Delta u_k + \cdots + b_{n-1} \Delta u_{k-n+1} \tag{3.14}$$

For convenience this can be separated into components which are known (depend on past inputs and outputs) and the d.o.f. (future input increments):

$$y_{k+1} = \underbrace{-[A_1, \cdots, A_{n+1}] \underset{\leftarrow k}{y} + [b_2, \cdots, b_n] \underset{\leftarrow k-1}{\Delta u}}_{\text{known}} + \underbrace{b_1 \Delta u_k}_{\text{d.o.f.}} \tag{3.15}$$

This equation gives a one-step ahead prediction for y_{k+1} given past data and the current input increment Δu_k. Hence this is analogous to equation (3.2).

3.3.1.2   Many step ahead prediction with $T(z) = 1$ – SISO case

An obvious way to compute predictions is by recursive use of (3.15), that is to use the same procedure as in Section 3.2 but based on one-step ahead prediction (3.15), i.e. compute y_{k+1} and substitute back into (3.15) at $k+2$ to get y_{k+2}, use the two-step ahead prediction to compute y_{k+3}, etc. Intuitively this appears straightforward, and is computationally very efficient. However, the algebra can be notationally awkward (see Section 3.4 for details) and hence is not the easiest procedure. There is a far more compact, *more insightful* and neater way of computing the predictions which is given next.

1. Write out the difference equation (3.14) for the next n_y sampling instants.

$$\begin{aligned}
y_{k+1} + A_1 y_k + \cdots + A_{n+1} y_{k-n} &= b_1 \Delta u_k + b_2 \Delta u_{k-1} + \cdots + b_n \Delta u_{k-n+1} \\
y_{k+2} + A_1 y_{k+1} + \cdots + A_{n+1} y_{k-n+1} &= b_1 \Delta u_{k+1} + b_2 \Delta u_k + \cdots + b_n \Delta u_{k-n+2} \\
&\vdots \\
y_{k+n_y} + \cdots + A_{n+1} y_{k+n_y+1-n} &= b_1 \Delta u_{k+n_y-1} + \cdots + b_n \Delta u_{k+n_y-n}
\end{aligned} \tag{3.16}$$

2. These can be placed in the following compact matrix/vector form

$$
\underbrace{\begin{bmatrix} 1 & 0 & \cdots & 0 \\ A_1 & 1 & \cdots & 0 \\ A_2 & A_1 & \cdots & 0 \\ \vdots & \vdots & \vdots & \vdots \end{bmatrix}}_{C_A} \underbrace{\begin{bmatrix} y_{k+1} \\ y_{k+2} \\ \vdots \\ y_{k+n_y} \end{bmatrix}}_{\underset{\rightarrow k}{y}} + \underbrace{\begin{bmatrix} A_1 & A_2 & \cdots & A_{n+1} \\ A_2 & A_3 & \cdots & 0 \\ A_3 & A_4 & \cdots & 0 \\ \vdots & \vdots & \vdots & \vdots \end{bmatrix}}_{H_A} \underbrace{\begin{bmatrix} y_k \\ y_{k-1} \\ \vdots \\ y_{k-n} \end{bmatrix}}_{\underset{\leftarrow k}{y}}
$$
$$
= \underbrace{\begin{bmatrix} b_1 & 0 & \cdots & 0 \\ b_2 & b_1 & \cdots & 0 \\ b_3 & b_2 & \cdots & 0 \\ \vdots & \vdots & \vdots & \vdots \end{bmatrix}}_{C_{zb}} \underbrace{\begin{bmatrix} \Delta u_k \\ \Delta u_{k+1} \\ \vdots \\ \Delta u_{k+n_y-1} \end{bmatrix}}_{\underset{\rightarrow k-1}{\Delta u}} + \underbrace{\begin{bmatrix} b_2 & b_3 & \cdots & b_n \\ b_3 & b_4 & \cdots & 0 \\ b_4 & b_5 & \cdots & 0 \\ \vdots & \vdots & \vdots & \vdots \end{bmatrix}}_{H_{zb}} \underbrace{\begin{bmatrix} \Delta u_{k-1} \\ \Delta u_{k-2} \\ \vdots \\ \Delta u_{k-n+1} \end{bmatrix}}_{\underset{\leftarrow k-1}{\Delta u}}
$$

(3.17)

3. Use the Toeplitz/Hankel notation (Section 1.4) to simplify (3.17) to

$$C_A \underset{\rightarrow k}{y} + H_A \underset{\leftarrow k}{y} = C_{zb} \underset{\rightarrow k-1}{\Delta u} + H_{zb} \underset{\leftarrow k-1}{\Delta u} \tag{3.18}$$

4. From (3.18) the output predictions are

$$\underset{\rightarrow k}{y} = C_A^{-1}[C_{zb}\underset{\rightarrow k-1}{\Delta u} + H_{zb}\underset{\leftarrow k-1}{\Delta u} - H_A \underset{\leftarrow k}{y}] \tag{3.19}$$

For convenience one may wish to represent (3.19) as

$$\underset{\rightarrow k}{y} = H\underset{\rightarrow k-1}{\Delta u} + P\underset{\leftarrow k-1}{\Delta u} + Q\underset{\leftarrow k}{y} \tag{3.20}$$

where $H = C_A^{-1}C_{zb}, P = C_A^{-1}H_{zb}$, $Q = -C_A^{-1}H_A$. C_A^{-1} can be computed very efficiently if required as $C_A^{-1} = C_{1/A}$. So one can simply use the coefficients of the
expansion $1/A(z)$. An efficient realisation of the matrix computations in (3.19) is
exactly equivalent to diophantine methods [102]. A worked example is given in Section A.1.2.

Summary: Output predictions, with $T = 1$, are given by

$$\underset{\rightarrow k}{y} = H\underset{\rightarrow k-1}{\Delta u} + P\underset{\leftarrow k-1}{\Delta u} + Q\underset{\leftarrow k}{y}$$
$$H = C_A^{-1}C_{zb}, \ P = C_A^{-1}H_{zb}, \ Q = -C_A^{-1}H_A \tag{3.21}$$

The predictions can be written down explicitly in terms of the model coefficients.

3.3.2 Prediction with CARIMA model and $T = 1$ – MIMO case

Assume here that a matrix fraction description model is going to be deployed. One
can reuse the development of the previous subsection without the need for any further

complication apart from the extra dimensionality. That is, the algebra has the same complexity as given in (3.16). In place of (3.14) use the difference equation

$$\mathbf{y}_{k+1} + D_1\mathbf{y}_k + \cdots + D_{n+1}\mathbf{y}_{k-n} = N_1\Delta\mathbf{u}_k + N_2\Delta\mathbf{u}_{k-1} + \cdots + N_n\Delta\mathbf{u}_{k-n+1} \qquad (3.22)$$

Then replicating all the steps of Section 3.3.1.2 one can jump straight to an equivalent matrix/vector form:

$$C_D\underset{\rightarrow k}{\mathbf{y}} + H_D\underset{\leftarrow k}{\mathbf{y}} = C_{zN}\underset{\rightarrow k-1}{\Delta\mathbf{u}} + C_{zN}\underset{\leftarrow k-1}{\Delta\mathbf{u}} \qquad (3.23)$$

Hence, by inspection

$$\underset{\rightarrow k}{\mathbf{y}} = H\underset{\rightarrow k-1}{\Delta\mathbf{u}} + P\underset{\leftarrow k-1}{\Delta\mathbf{u}} + Q\underset{\leftarrow k}{\mathbf{y}} \qquad (3.24)$$

where $H = C_D^{-1}C_{zN}$, $P = C_D^{-1}H_{zN}$, $Q = -C_D^{-1}H_D$.

Summary: The algebraic complexity for the MIMO case is identical to the SISO case. Predictions are given as

$$\underset{\rightarrow k}{\mathbf{y}} = H\underset{\rightarrow k-1}{\Delta\mathbf{u}} + P\underset{\leftarrow k-1}{\Delta\mathbf{u}} + Q\underset{\leftarrow k}{\mathbf{y}} \qquad (3.25)$$

3.3.3 Prediction equations with $T(z) \neq 1$

It has been noted by many authors that the T-filter [21] is often essential for practical applications of MPC based on transfer function models[*]. So the following development is important.

To skip straight to the solution go to eqn.(3.34) where the analogy with eqn.(3.21) can be seen.

Before proceeding with the details remind yourself of the rules for algebra of Toeplitz and Hankel matrices (Sections 1.4.3.3, 1.4.3.5). Also here we treat only the SISO case as the algebra for the MIMO and SISO case is identical when one uses the Toeplitz/Hankel notation for computing predictions.

3.3.3.1 Summary of the key steps in computing prediction equations

Including a T-filter the CARIMA model is given as

$$Ay_k = b(z)\Delta u_k + Tv_k \qquad (3.26)$$

In the earlier sections, the effect of the term $T(z)v(z)$ was ignored or assumed to have an expected value of zero. Of course this is a conservative assumption as some of the

---
[*]This is not so with FIR models and independent models e.g. [127].

past values of v_k can be inferred. One can replace (3.26) by an alternative in which the unknown term is zero mean and unknown simply by filtering left- and right-hand sides by $T(z)$, i.e.

$$\frac{A}{T} y_k = \frac{b}{T} \Delta u_k + v_k \qquad (3.27)$$

If one uses (3.27) for prediction, then the bias due to past values of v_k is removed hence improving prediction accuracy.

> **Summary:** The T-filter improves prediction accuracy by using model (3.27) for prediction in place of (3.12).

Rearrange (3.27) as follows:

$$A[\frac{y_k}{T}] = b[\frac{\Delta u_k}{T}] + v_k \quad \Rightarrow \quad A\tilde{y}_k = b\Delta\tilde{u}_k + v_k \qquad (3.28)$$

where filtered variables are defined as

$$\tilde{y}_k = \frac{y_k}{T}; \quad \Delta\tilde{u}_k = \frac{\Delta u_k}{T} \qquad (3.29)$$

In this way the intuitive interpretation of the T-filter is that it is used to filter input/output (I/O) data before prediction. Typically $1/T(z)$ is a low-pass filter so the filtering reduces the effects of high frequency noise.

The benefits of the filtering are obtained by using (3.28) for prediction in place of (3.12), the only difference being the substitution of filtered signals for unfiltered signals. Hence, by inspection (from eqn.(3.21)) one can write the prediction equation for filtered variables as

$$\underset{\rightarrow k}{\tilde{y}} = H\Delta\underset{\rightarrow k-1}{\tilde{u}} + P\Delta\underset{\leftarrow k-1}{\tilde{u}} + Q\underset{\leftarrow k}{\tilde{y}} \qquad (3.30)$$

Unfortunately this is no use because we want predictions for $\underset{\rightarrow k}{y}$ in terms of $\Delta\underset{\rightarrow k-1}{u}$; that is, future signals must be unfiltered.

> **Summary:**
>
> 1. The T-filter is usually a low-pass filter and hence improves prediction accuracy in the high frequency range by reducing the transference of high frequency noise.
>
> 2. Use of (3.28, 3.30) gives predictions in terms of future filtered variables.

3.3.3.2 Forming the prediction equations with a T-filter using predictions (3.21)

One can easily reconstitute values for future unfiltered variables by writing the filter equations (3.29) in Toeplitz/Hankel form:

$$C_T\underset{\rightarrow}{\tilde{y}} + H_T\underset{\leftarrow}{\tilde{y}} = \underset{\rightarrow}{y}; \quad C_T\Delta\underset{\rightarrow}{\tilde{u}} + H_T\Delta\underset{\leftarrow}{\tilde{u}} = \Delta\underset{\rightarrow}{u} \qquad (3.31)$$

Substituting (3.31) into (3.30) gives:

$$\underbrace{C_T^{-1}[\underset{\rightarrow}{y} - H_T\underset{\leftarrow}{\tilde{y}}]}_{\underset{\rightarrow k}{\tilde{y}}} = H\underbrace{C_T^{-1}[\underset{\rightarrow}{\Delta u} - H_T\Delta\tilde{u}]}_{\underset{\rightarrow}{\Delta\tilde{u}}} + P\Delta\underset{\leftarrow}{\tilde{u}} + Q\underset{\leftarrow}{\tilde{y}} \qquad (3.32)$$

Multiplying left and right by C_T and then grouping common terms gives

$$\underset{\rightarrow k}{y} = H\Delta\underset{\rightarrow k-1}{u} + \tilde{P}\Delta\underset{\leftarrow k-1}{\tilde{u}} + \tilde{Q}\underset{\leftarrow k}{\tilde{y}}; \qquad \begin{cases} \tilde{P} = [C_T P - HH_T] \\ \tilde{Q} = [H_T + C_T Q] \end{cases} \qquad (3.33)$$

Derivation of prediction equations is very similar with and without $T(z)$. The differences when $T \neq 1$ are

1. Filtering on past data is used.

2. Matrices are changed as: $P \Rightarrow \tilde{P}, \ Q \Rightarrow \tilde{Q}$.

3. If P, Q are known, then $\tilde{P}, \tilde{Q}$ (see eqn. 3.34) can be computed very efficiently from (3.33).

4. If P, Q are not known, use eqn. (3.40) derived below.

5. If $T = 1$, then $H_T = 0$, $C_T = I$ so $P = \tilde{P}$, $Q = \tilde{Q}$.

Summary: The prediction equations with a T-filter are based on filtered values of past data through

$$\underset{\rightarrow k}{y} = H\Delta\underset{\rightarrow k-1}{u} + \tilde{P}\Delta\underset{\leftarrow k-1}{\tilde{u}} + \tilde{Q}\underset{\leftarrow k}{\tilde{y}} \quad \text{where} \quad \tilde{y} = y/T, \quad \Delta\tilde{u} = \Delta u/T \qquad (3.34)$$

Matrices $\tilde{P}, \tilde{Q}$ are related to P, Q of (3.21) through

$$\tilde{P} = [C_T P - HH_T]; \quad \tilde{Q} = [H_T + C_T Q] \qquad (3.35)$$

3.3.3.3 Prediction equation derivation based directly on model parameters

If one were to assume that H, P, Q were unknown and hence prediction (3.21) is unknown, then one can derive prediction equation (3.34) directly from (3.27) and (3.31). The development is given below. For simplicity the sample time subscript is dropped.

From (3.18) the prediction equation for (3.27) is

$$C_A\underset{\rightarrow}{y} + H_A\underset{\leftarrow}{y} = C_{zb}\Delta\underset{\rightarrow}{u} + H_{zb}\Delta\underset{\leftarrow}{u} \qquad (3.36)$$

Substituting in from (3.31) gives

$$C_A C_T^{-1}[\underset{\rightarrow}{y} - H_T\underset{\leftarrow}{\tilde{y}}] + H_A\underset{\leftarrow}{\tilde{y}} = C_b C_T^{-1}[\Delta\underset{\rightarrow}{u} - H_T\Delta\underset{\leftarrow}{\tilde{u}}] + H_b\Delta\underset{\leftarrow}{\tilde{u}} \qquad (3.37)$$

Toeplitz matrices of SISO functions commute (e.g. $C_A C_T^{-1} = C_T^{-1} C_A$), therefore eqn. (3.37) can be rearranged as follows:

$$C_T^{-1} C_A \underset{\rightarrow}{y} - C_T^{-1} C_A H_T \underset{\leftarrow}{\tilde{y}} + H_A \underset{\leftarrow}{\tilde{y}} = C_T^{-1} C_{zb} \underset{\rightarrow}{\Delta u} - C_T^{-1} C_{zb} H_T \underset{\leftarrow}{\Delta \tilde{u}} + H_{zb} \underset{\leftarrow}{\Delta \tilde{u}} \qquad (3.38)$$

Now, (i) move known terms to RHS, (ii) multiply by C_T, (iii) remove cancelling factors, (iv) multiply by C_A^{-1}:

$$
\begin{aligned}
C_T^{-1} C_A \underset{\rightarrow}{y} &= C_T^{-1} C_{zb} \underset{\rightarrow}{\Delta u} + [H_{zb} - C_T^{-1} C_{zb} H_T] \underset{\leftarrow}{\Delta \tilde{u}} + [C_T^{-1} C_A H_T - H_A] \underset{\leftarrow}{\tilde{y}} \\
C_A \underset{\rightarrow}{y} &= C_T C_T^{-1} C_{zb} \underset{\rightarrow}{\Delta u} + C_T [H_{zb} - C_T^{-1} C_{zb} H_T] \underset{\leftarrow}{\Delta \tilde{u}} + C_T [C_T^{-1} C_A H_T - H_A] \underset{\leftarrow}{\tilde{y}} \\
C_A \underset{\rightarrow}{y} &= C_{zb} \underset{\rightarrow}{\Delta u} + [C_T H_{zb} - C_{zb} H_T] \underset{\leftarrow}{\Delta \tilde{u}} + [C_A H_T - C_T H_A] \underset{\leftarrow}{\tilde{y}} \\
\underset{\rightarrow}{y} &= C_A^{-1} \{ C_{zb} \underset{\rightarrow}{\Delta u} + [C_T H_{zb} - C_{zb} H_T] \underset{\leftarrow}{\Delta \tilde{u}} + [C_A H_T - C_T H_A] \underset{\leftarrow}{\tilde{y}} \}
\end{aligned}
$$
$$(3.39)$$

The predictions are now in the form of (3.34)

$$\underset{\rightarrow}{y} = H \underset{\rightarrow}{\Delta u} + \tilde{P} \underset{\leftarrow}{\Delta \tilde{u}} + \tilde{Q} \underset{\leftarrow}{\tilde{y}}; \qquad \left\{ \begin{aligned} H &= C_A^{-1} C_{zb} \\ \tilde{P} &= C_A^{-1} [C_T H_{zb} - C_{zb} H_T] \\ \tilde{Q} &= C_A^{-1} [C_A H_T - C_T H_A] \end{aligned} \right. \qquad (3.40)$$

Summary: The predictions with the T-filter, eqn.(3.40), can be written explicity in terms of the model parameters.

3.4 Using recursion to find matrices H, P, Q

This section is here solely because it gives an efficient means of computing the predictions for coding purposes. You are advised to *skip it* unless computational efficiency is more important than having compact equations. The solutions given are identical to those that arise from diophantine methods (e.g. [21]). Although it complicates notation a little, for generality the MIMO case will be given here with an MFD model.

1. Assume an underlying difference equation for the one-step ahead prediction

$$\mathbf{y}_{k+1} + D_1 \mathbf{y}_k + \cdots + D_{n+1} \mathbf{y}_{k-n} = N_1 \Delta \mathbf{u}_k + N_2 \Delta \mathbf{u}_{k-1} + \cdots + N_n \Delta \mathbf{u}_{k-n+1} \quad (3.41)$$

2. Introduce notation for prediction horizon $(.)^{[i]}$ to denote i-step ahead prediction such that in general

$$\mathbf{y}_{k+i} = H^{[i]} \underset{\rightarrow}{\Delta \mathbf{u}}_{k-1} + P^{[i]} \underset{\leftarrow}{\Delta \mathbf{u}}_{k-1} + Q^{[i]} \underset{\leftarrow}{\mathbf{y}}_k \qquad (3.42)$$

3. Initialise (3.42) for $i = 1$ from (3.41)

$$H^{[1]} = [N_1, 0, 0, 0, \ldots]; \quad P^{[1]} = [N_2, N_3, \ldots, N_n]; \quad Q^{[1]} = [-D_1, -D_2, \ldots, -D_{n+1}]$$
(3.43)

4. Use recursive substitution to find $\mathbf{y}_{k+i+1}$ in terms of $\mathbf{y}_{k+i}$ and $\mathbf{y}_{k+1}$.

   (a) Given the coefficients of the prediction equation for $\mathbf{y}_{k+i}$, compute $\mathbf{y}_{k+i+1}$ as follows:

$$\mathbf{y}_{k+i} = H^{[i]} \Delta \mathbf{u}_{\underset{\rightarrow}{k-1}} + P^{[i]} \Delta \mathbf{u}_{\underset{\leftarrow}{k-1}} + Q^{[i]} \mathbf{y}_{\underset{\leftarrow}{k}}$$

$$\Downarrow$$
(3.44)

$$\mathbf{y}_{k+i+1} = H^{[i]} \Delta \mathbf{u}_{\underset{\rightarrow}{k}} + P^{[i]} \Delta \mathbf{u}_{\underset{\leftarrow}{k}} + Q^{[i]} \mathbf{y}_{\underset{\leftarrow}{k+1}}$$

   (b) Note that

$$Q^{[i]} \mathbf{y}_{\underset{\leftarrow}{k+1}} = [Q^{[i]}, 0] \begin{bmatrix} \mathbf{y}_{k+1} \\ \mathbf{y}_{\underset{\leftarrow}{k}} \end{bmatrix} = Q_1^{[i]} \mathbf{y}_{k+1} + [Q_2^{[i]}, \ldots, Q_n^{[i]}, 0] \mathbf{y}_{\underset{\leftarrow}{k}}$$

$$P^{[i]} \Delta \mathbf{u}_{\underset{\leftarrow}{k}} = [P^{[i]}, 0] \begin{bmatrix} \Delta \mathbf{u}_k \\ \Delta \mathbf{u}_{\underset{\leftarrow}{k-1}} \end{bmatrix} = P_1^{[i]} \Delta \mathbf{u}_k + [P_2^{[i]}, \ldots, P_{n-1}^{[i]}, 0] \Delta \mathbf{u}_{\underset{\leftarrow}{k-1}}$$
(3.45)

   (c) Substitute observations (3.45) into prediction (3.44b)

$$\mathbf{y}_{k+i+1} = [P_1^{[i]}, H^{[i]}] \Delta \mathbf{u}_{\underset{\rightarrow}{k-1}} + [P_2^{[i]}, \ldots, P_{n-1}^{[i]}, 0] \Delta \mathbf{u}_{\underset{\leftarrow}{k-1}}$$
$$+ [Q_2^{[i]}, \ldots, Q_n^{[i]}, 0] \mathbf{y}_{\underset{\leftarrow}{k}} + Q_1^{[i]} \mathbf{y}_{k+1}$$
(3.46)

   (d) Substitute $\mathbf{y}_{k+1}$ from (3.42) into (3.46), to give

$$\mathbf{y}_{k+i+1} = [P_1^{[i]}, H^{[i]}] \Delta \mathbf{u}_{\underset{\rightarrow}{k-1}} + [P_2^{[i]}, \ldots, P_{n-1}^{[i]}, 0] \Delta \mathbf{u}_{\underset{\leftarrow}{k-1}} + [Q_2^{[i]}, \ldots, Q_n^{[i]}, 0] \mathbf{y}_{\underset{\leftarrow}{k}}$$
$$+ Q_1^{[i]} [H^{[1]} \Delta \mathbf{u}_{\underset{\rightarrow}{k-1}} + P^{[1]} \Delta \mathbf{u}_{\underset{\leftarrow}{k-1}} + Q^{[1]} \mathbf{y}_{\underset{\leftarrow}{k}}]$$
(3.47)

   (e) Finally, group common terms to give the form of (3.42) again

$$\mathbf{y}_{k+i+1} = \underbrace{[P_1^{[i]} + Q_1^{[i]} H_1^{[1]}, H^{[i]}]}_{H^{[i+1]}} \Delta \mathbf{u}_{\underset{\rightarrow}{k-1}}$$
$$+ \underbrace{\{[P_2^{[i]}, \ldots, P_{n-1}^{[i]}, 0] + Q_1^{[i]} P^{[1]}\}}_{P^{[i+1]}} \Delta \mathbf{u}_{\underset{\leftarrow}{k-1}}$$
(3.48)
$$+ \underbrace{\{[Q_2^{[i]}, \ldots, Q_n^{[i]}, 0] + Q_1^{[i]} Q^{[1]}\}}_{Q^{[i+1]}} \mathbf{y}_{\underset{\leftarrow}{k}}$$

Remark 3.2 *Recursion (3.48) has the same computational complexity of the matrix approach given earlier. However it may be preferable in packages that do not support matrix algebra.*

Remark 3.3 *The overall prediction equation analogous to (3.21) is determined using*

$$
H = \begin{bmatrix} H^{[1]}\,0\,0\,0\ldots \\ H^{[2]}\,0\,0\ldots \\ H^{[3]}\,0\ldots \\ \vdots\;\;\vdots\;\;\vdots\;\;\vdots \end{bmatrix}; \quad P = \begin{bmatrix} P^{[1]} \\ P^{[2]} \\ P^{[3]} \\ \vdots \end{bmatrix}; \quad Q = \begin{bmatrix} Q^{[1]} \\ Q^{[2]} \\ Q^{[3]} \\ \vdots \end{bmatrix} \tag{3.49}
$$

Note that $H^{[i+1]}$ has one more nonzero column than $H^{[i]}$. This was clear from the earlier observation that H is a Toeplitz matrix $C_{G/\Delta}$ of the step response.

Summary: Given a generic prediction equation

$$
\mathbf{y}_{k+i} = H^{[i]}\Delta\underrightarrow{\mathbf{u}}_{k-1} + P^{[i]}\Delta\underleftarrow{\mathbf{u}}_{k-1} + Q^{[1]}\underleftarrow{\mathbf{y}}_k \tag{3.50}
$$

One can initialise with $i = 1$ and find the remaining prediction coefficients from the recursion

$$
\begin{aligned}
H^{[i+1]} &= [P_1^{[i]} + Q_1^{[i]}H_1^{[1]}, H^{[i]}] \\
P^{[i+1]} &= [P_2^{[i]}, ..., P_{n-1}^{[i]}, 0] + Q_1^{[i]}P^{[1]} \\
Q^{[i+1]} &= [Q_2^{[i]}, ..., Q_n^{[i]}, 0] + Q_1^{[i]}Q^{[1]}
\end{aligned} \tag{3.51}
$$

3.5 Prediction with FIR models

FIR models are the most common models utilised in commerical MPC packages. There is a good reason for this which will be discussed later. They are restricted to stable processes. For now let us concentrate solely on how these models are used for prediction. As with the other models, some care must be taken to ensure offset free prediction in the steady-state case. Let the underlying model be

$$
y(z) = G(z)u(z) + \frac{v}{\Delta(z)}; \quad G(z) = \sum_{i=0}^{\infty} G_i z^{-1} \tag{3.52}
$$

where v is unknown. (This is equivalent to the CARIMA model with $A = T$.)

3.5.1 Impulse response models

These are not commonly used so are mentioned here solely for completeness. Ignoring the disturbance, the model prediction for the current sample is

$$
\hat{y}_k = \sum_{i=0}^{\infty} G_i u_{k-i} \tag{3.53}
$$

Define the offset between model output $\hat{y}_k$ and current measurement y_k:

$$\hat{d}_k = y_k - \hat{y}_k \tag{3.54}$$

Assume this offset represents the best estimate of the unknown term $d_k = \frac{v_k}{\Delta}$. Hence the process output predictions can be derived from

$$\hat{y}_{k+j} = \sum_{i=0}^{\infty} G_i u_{k+j-i} + \hat{d}_k \tag{3.55}$$

The algebra is far easier than with state-space models and transfer function models because there is no auto-regressive part in the model; predictions depend solely on the input information. The downside of this is that the model order has to be very high to avoid large bias errors. That is, in practice one will use

$$\hat{y}_{k+j} = y_k + \sum_{i=0}^{n} G_i u_{k+j-i} - \underbrace{\sum_{i=0}^{n} G_i u_{k-i}}_{\hat{y}_k} \tag{3.56}$$

and n must be large enough such that $i > n \implies |G_i|$ negligible.

Remark 3.4 *Offset free prediction is incorporated via the disturbance estimate $\hat{d}_k$. That is, assume $\mathbf{u}_{k+i} = \mathbf{u}_k = \mathbf{u}_{k-1} + \Delta\mathbf{u}_k$. One can then compute $\Delta\mathbf{u}_k$ to ensure that $E[\mathbf{y}_{k+i}] = \mathbf{r}, \forall i > 0$ and in steady state this will be consistent iff $\Delta\mathbf{u}_k = 0$.*

3.5.2   Step response models

Prediction follows a similar technique to that for impulse response FIR. First write the model as

$$y_k = \frac{G(z)}{\Delta(z)} \Delta u_k + \frac{v_k}{\Delta(z)} \tag{3.57}$$

where one notes that the equation contains input increments (changes in control). Let $G/\Delta = \sum_{i=0}^{\infty} H_i z^{-i}$. The predictions are given as

$$\hat{y}_{k+i} = d_k + \sum_{j=0}^{\infty} H_j \Delta u_{k+i-j|k} \tag{3.58}$$

With y_k the most recent measurement, the best estimate of d_k is

$$y_k - \sum_{j=0}^{\infty} H_j \Delta u_{k-j|k} = \hat{d}_k \tag{3.59}$$

Hence eliminating the unknown d_k from (3.58) and using an arbitrary truncation n give the prediction

$$y_{k+i} = y_k + \sum_{j=0}^{i-1} H_{i-j} \Delta u_{k+j} + \sum_{j=1}^{n} [H_{j+i} - H_j] \Delta u_{k-j} \tag{3.60}$$

This is similar in structure to that of (3.56) except that here the neglected terms are $[H_{n+i} - H_n]$ and one must choose n large enough (theoretically $n = \infty$ but it is known that $H_{j+i} - H_j \to 0$ for large j) so that these are small enough.

Summary: For a step response model, the prediction takes the form

$$\underset{\to k}{y} = H \underset{\to k-1}{\Delta u} + L\hat{y}_k + M \underset{\leftarrow k-1}{\Delta u} \qquad (3.61)$$

where C_H is the Toeplitz matrix of $H(z)$ and

$$H = C_{G/\Delta}; \quad M = \begin{bmatrix} H_2 - H_1 & H_3 - H_2 & H_4 - H_3 & \dots & H_{n+1} - H_n \\ H_3 - H_1 & H_4 - H_2 & H_5 - H_2 & \dots & H_{n+2} - H_n \\ \vdots & \vdots & \vdots & \vdots & \vdots \end{bmatrix}; \quad L = \begin{bmatrix} I \\ I \\ \vdots \end{bmatrix} \qquad (3.62)$$

3.6 Prediction with independent models

As mentioned in Section 2.6, independent models (IM or internal models) are simulated in parallel with the process. The hope is that the IM will give a similar output to the real process. Clearly in practice the two will differ due to the disturbances and model uncertainty.

3.6.1 Structure and prediction set up with internal models

When using IM for prediction, it is equivalent to the use of an FIR model with no truncation errors; this is because the prediction is based solely on input information. The internal state of the IM is determined solely by past inputs (and an initial condition whose effect in general will have decayed to zero). The advantage of IM over FIR models is that the prediction can be made using equations (3.10) or (3.21) instead of (3.56) or (3.62) and these involve far fewer parameters in general.

To illustrate, consider the scenario given in Figure 3.1.

1. **IM a state-space model:** At a given sampling instant this model has a known state $\check{x}_k$, which need not match the unknown process state x_k (recall there is no observer because the IM state is derived from a a computer simulation). From (3.10) and the offset correction (3.54) the system predictions are given by

$$\underset{\to k}{y} = P_x \check{x}_k + H \underset{\to k-1}{\Delta u} + [y_k - \check{y}_k] \qquad (3.63)$$

2. **IM an MFD model:** Use predictions (3.21) for the model and offset correction

(3.54) to give

$$\underset{\rightarrow k}{\mathbf{y}} = H\Delta\underset{\rightarrow k-1}{\mathbf{u}} + P\Delta\underset{\leftarrow k-1}{\mathbf{u}} + Q\underset{\leftarrow k}{\check{\mathbf{y}}} + L[\mathbf{y}_k - \check{\mathbf{y}}_k] \qquad (3.64)$$

It is noted with an IM there is no need for a T-filter, as one is using a model equivalent to (3.52).

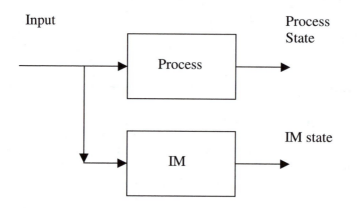

FIGURE 3.1
Independent model state and process state.

3.6.2   Prediction with unstable open-loop plant

It was noted in passing that one cannot use FIR models for open-loop unstable plant, because the sequence models would diverge. One might think that such a weakness also carries over to IM in that they have an equivalence to IM models and are simulated open-loop. The state of the IM would diverge because an input sequence stabilising the plant is unlikely to stabilise the IM which practically, due to uncertainty, will be different.

3.6.2.1   Decomposition for IM

Practitioners who use IM for commerical packages (e.g. ADERSA) have produced a simple mechanism for getting around this. Here an illustration is given for the SISO case only. First decompose the system as follows:

$$G = \frac{M_1}{1 - M_2} \tag{3.65}$$

where M_1, M_2 are both stable transfer functions. Such a decomposed system can be represented by the block diagram of Figure 3.2 where y_m is the output of the model. Simulating Figure 3.2 would be identical to simulating G and gives no benefit. However, postulate that the output y of the process should match that of the IM (that is $y = y_m$); in this case one can replace the positive feedback loop by a simple transference from the plant output, as in Figure 3.3. Assuming that the process is stabilised, then the IM output y_m must also be stable because by definition both M_1, M_2 are stable.

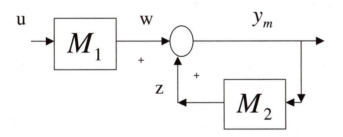

FIGURE 3.2
Block diagram of IM decomposition for unstable open-loop plant.

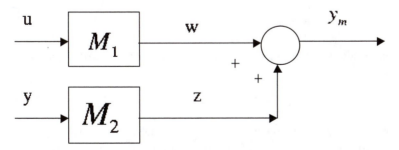

FIGURE 3.3
Block diagram of IM simulation for unstable open-loop plant.

3.6.2.2 Selection of decomposition

The best way to decompose is still an open question. A convenient means which is assumed hereafter is given next.

$$G = \frac{n}{d} = \frac{n_+ n_-}{d_+ d_-}; \quad M_1 = \frac{n_+}{d_-}; \quad M_2 = \frac{b_2}{n_-}; \quad b_2 = n_- - d_+ \qquad (3.66)$$

where n_+, d_+ are the numerator/denominator factors, respectively, containing roots outside the unit circle (unstable). It is clear that both M_1 and M_2 have stable poles.

3.6.2.3 Prediction using decomposed IM

For those who do not like algebra, go straight to the result (3.75) which is relatively easy to code.

Actually prediction with the IM represented in Figures 3.2, 3.3 is not simple. One would only go this route if one thought that there are significant benefits from using an IM as opposed to say a more typical state-space model with an observer or a transfer function model with a T-filter. The benefits of IM are clear for stable processes [127], but although expected have not been clearly demonstrated in this case. Further discussion of this will be given in a later section on sensitivity.

First establish how an IM is used on a real application.

- The open-loop simulation deploys Figure 3.3 and hence defines the variables w and z.

- Prediction is based on Figure 3.2 (as future y are unknown hence Figure 3.3 cannot be used). The values of w, z in Figure 3.2 are re-intialised at each sampling instant from those in Figure 3.3.

Set up prediction equations around M_1 and M_2 separately:

$$\begin{aligned}
C_{d_-} \underset{\rightarrow}{w} &= C_{n_+} \underset{\rightarrow}{u} + H_{n_+} \underset{\leftarrow}{u} - H_{d_-} \underset{\leftarrow}{w} \\
C_{n_-} \underset{\rightarrow}{z} &= C_{b_2}[\underset{\rightarrow}{y_m} + \hat{d}] + H_{b_2} \underset{\leftarrow}{y} - H_{n_-} \underset{\leftarrow}{z} \\
\underset{\rightarrow}{y_m} &= \underset{\rightarrow}{z} + \underset{\rightarrow}{w} \\
\underset{\rightarrow}{y} &= \underset{\rightarrow}{y_m} + \hat{d}
\end{aligned} \qquad (3.67)$$

where $\hat{d}$ represents a correction for offset which is used in IM based MPC (and DMC etc.) to ensure integral action. Define

$$\hat{d} = L(\underset{\leftarrow}{y} - \underset{\leftarrow}{z} - \underset{\leftarrow}{w}) \qquad (3.68)$$

where L is a vector of ones. For convenience (to ensure that $J = 0$ is consistent with zero offset) it is usual to express the d.o.f. in terms of future input increments (rather than absolute inputs), hence

$$\underset{\rightarrow}{u} = E\Delta \underset{\rightarrow}{u} + L\underset{\leftarrow}{u} \qquad (3.69)$$

where $E = C_{1/\Delta}$ is a lower triangular matrix of ones. One can then rewrite (3.67) as

$$
\begin{aligned}
C_{d_-}\underrightarrow{w} &= C_{n_+}(E\Delta\underrightarrow{u} + L\underleftarrow{u}) + H_{n_+}\underleftarrow{u} - H_{d_-}\underleftarrow{w} \\
C_{n_-}\underrightarrow{z} &= C_{b_2}[\underrightarrow{z} + \underrightarrow{w} + L(\underleftarrow{y} - \underleftarrow{z} - \underleftarrow{w})] + H_{b_2}\underleftarrow{y} - H_{n_-}\underleftarrow{z} \\
\underrightarrow{y_m} &= \underrightarrow{z} + \underrightarrow{w} \\
\underrightarrow{y} &= \underrightarrow{y_m} + L(\underleftarrow{y} - \underleftarrow{z} - \underleftarrow{w})
\end{aligned}
\tag{3.70}
$$

Summary: With an IM of an unstable process, open-loop prediction reduces to the solution of the simultaneous equations in (3.68, 3.70).

3.6.2.4 Algebraic solution for prediction equations (3.70)

In the first instance, solve the simultaneous equations (3.70) for $\underrightarrow{y_m}$:

$$
[I - C_{n_-}^{-1}C_{b_2}]\underrightarrow{y_m} = \{\; C_d^{-1}[C_{n_+}(E\Delta\underrightarrow{u} + L\underleftarrow{u}) + H_{n_+}\underleftarrow{u} - H_{d_-}\underleftarrow{w}] \\
+ C_{n_-}^{-1}[C_{b_2}L(\underleftarrow{y} - \underleftarrow{z} - \underleftarrow{w}) + H_{b_2}\underleftarrow{y} - H_{n_-}\underleftarrow{z}]\}
\tag{3.71}
$$

However, note that $[I - C_{n_-}^{-1}C_{b_2}] = C_{d_+}C_{n_-}^{-1}$ and hence:

$$
\begin{aligned}
\underrightarrow{y_m} &= C_{d_+}^{-1}C_{n_-}\{\; C_d^{-1}[C_{n_+}(E\Delta\underrightarrow{u} + L\underleftarrow{u}) + H_{n_+}\underleftarrow{u} - H_{d_-}\underleftarrow{w}] \\
&\quad + C_{n_-}^{-1}[C_{b_2}L(\underleftarrow{y} - \underleftarrow{z} - \underleftarrow{w}) + H_{b_2}\underleftarrow{y} - H_{n_-}\underleftarrow{z}]\} \\
&= C_{d_+}^{-1}C_d^{-1}\{\; C_{n_-}[C_{n_+}(E\Delta\underrightarrow{u} + L\underleftarrow{u}) + H_{n_+}\underleftarrow{u} - H_{d_-}\underleftarrow{w}] \\
&\quad + C_{d_-}[C_{b_2}L(\underleftarrow{y} - \underleftarrow{z} - \underleftarrow{w}) + H_{b_2}\underleftarrow{y} - H_{n_-}\underleftarrow{z}]\}
\end{aligned}
\tag{3.72}
$$

Putting common terms together gives

$$
\underrightarrow{y_m} = C_d^{-1}\{\; C_{n_-}[C_{n_+}E\Delta\underrightarrow{u} + (C_{n_+}L + H_{n_+})\underleftarrow{u}] \\
- (C_{n_-}H_{d_-} + C_{d_-}C_{b_2}L)\underleftarrow{w} + C_{d_-}(C_{b_2}L + H_{b_2})\underleftarrow{y} - C_{d_-}(C_{b_2}L + H_{n_-})\underleftarrow{z}\}
\tag{3.73}
$$

Finally adding in the equation $\underrightarrow{y} = \underrightarrow{y_m} + L(\underleftarrow{y} - \underleftarrow{z} - \underleftarrow{w})$ and tidying up gives

$$
\begin{aligned}
\underrightarrow{y} &= H\Delta\underrightarrow{u} + P_u\underleftarrow{u} + P_w\underleftarrow{w} + P_y\underleftarrow{y} + P_z\underleftarrow{z} \\
H &= C_d^{-1}C_nE \\
P_u &= C_d^{-1}(C_nL + C_{n_-}H_{n_+}) \\
P_w &= -C_d^{-1}(C_{n_-}H_{d_-} + C_{d_-}C_{b_2}L) - L \\
P_y &= C_d^{-1}C_{d_-}(C_{b_2}L + H_{b_2}) + L \\
P_z &= -C_d^{-1}C_{d_-}(C_{b_2}L + H_{n_-}) - L
\end{aligned}
\tag{3.74}
$$

Summary: The prediction equations for Figures 3.2, 3.3 can be summarised in a neat form corresponding to (3.1), that is

$$\underrightarrow{y} = H\Delta\underrightarrow{u} + Mv; \qquad v = \begin{bmatrix} \underleftarrow{u} \\ \underleftarrow{w} \\ \underleftarrow{y} \\ \underleftarrow{z} \end{bmatrix} \qquad (3.75)$$

$$M = [P_u, P_w, P_y, P_z];$$

Clearly there is a nice separation between the part $H\Delta\underrightarrow{u}$ depending on the d.o.f. (often called 'forced response') and the notional 'free response' part Mv, i.e. the response should the input remain unchanged.

3.7 Numerically robust prediction with open-loop unstable plant

This section takes its content from two main publications [104, 113]. Some illustration of this is given in Section 7.5.2. The key warnings follow.

Warnings:

- Open-loop prediction of unstable processes often leads to poor numerical conditioning of MPC problems.

- Poor conditioning can result in arbitrary values and at best severely suboptimal values for the control law coefficients.

- If you have an unstable open-loop process, analyse your controller and intermediate computations carefully.

The proposal given here is a simple way of reducing numerical ill-conditioning. If the system is open-loop unstable, then you must use pseudo closed-loop predictions; that is, prestabilise the plant before predicting. The choice of prestabilisation is not critical though different choices have different advantages. The mechanics/algebra for this is also fairly straightforward and will be explained next.

Key message: Use prestabilised predictions to ensure correct calculation of the control law.

3.7.1 Why is open-loop prediction unsatisfactory?

With an unstable process, the open-loop step response is divergent. Hence consider the matrix H which comprises the Toeplitz matrix of the step response. This will contain some very small numbers in the upper rows and some very large numbers in the lower rows. The end result is a matrix with worsening conditioning as the horizons increase and notably it is usual in MPC to deploy large horizons.

Due to computer rounding, manipulations of H would always create further round off errors. For a well-conditioned matrix this is unimportant. However, if a matrix is poorly conditioned, these round-off errors can make a large difference to the result. If matrices H, P, Q, P_x are poorly conditioned, they may not be reliable for use in an MPC control law. This is particularly true of H for which a typical manipulation is

$$M = (H^T H + I)^{-1} H^T \qquad (3.76)$$

So if H has very large coefficients and poor conditioning, the matrix $(H^T H + I)$ may become near rank deficient to within computer accuracy and cannot be inverted reliably.

Summary: If you have an open-loop unstable plant and insist on using open-loop predictions, either: (i) use small horizons or (ii) check very carefully the conditioning of your calculations.

3.7.2 Prestabilisation and pseudo closed-loop prediction

Although there are many alternative ways of prestabilising, here we will illustrate briefly just two. Further discussion is given in Chapter 7.

1. Select the future control trajectory $\Delta \underrightarrow{u}$ to cancel any unstable poles. ([42, 94, 107]).

2. Choose a stabilising feedback and pseudo closed-loop prediction.

For those concerned about the terminology *pole cancellation*, this is done in a receding horizon sense (via feedback) and so is not vunerable to small errors due to model uncertainty. In this section, illustrations are given of how this can be done.

3.7.2.1 Prestabilisation by pole cancellation

Take for example purposes the prediction equations (3.19).

1. Decompose the open-loop poles as

$$A(z) = A_+(z)A_-(z) \qquad (3.77)$$

where A_+ contains all the unstable poles.

2. Rewrite the prediction terms of the equivalent z-transforms using:

$$\underrightarrow{y}(z) = [1 \; z^{-1} \; z^{-2} \; \ldots]\underrightarrow{y}; \quad \Delta \underrightarrow{u}(z) = [1 \; z^{-1} \; z^{-2} \; \ldots]\Delta \underrightarrow{u} \qquad (3.78)$$

Note that this implicitly gives the predictions for an infinite horizon.

3. Take the equation

$$C_A \underset{\rightarrow}{y} = C_b \underset{\rightarrow}{\Delta u} + H_b \underset{\leftarrow}{\Delta u} - H_A \underset{\leftarrow}{\Delta u} \tag{3.79}$$

In transfer function form this is equivalent to

$$\underset{\rightarrow}{y}(z) = \frac{[1 \; z^{-1} \; z^{-2} \; \dots](C_b \underset{\rightarrow}{\Delta u} + H_b \underset{\leftarrow}{\Delta u} - H_A \underset{\leftarrow}{\Delta u})}{A(z)} = \frac{b(z)\underset{\rightarrow}{\Delta u}(z) + p(z)}{A(z)} \tag{3.80}$$

where $p = [1 \; z^{-1} \; z^{-2} \; \dots](H_b \underset{\leftarrow}{\Delta u} - H_A \underset{\leftarrow}{\Delta u})$.

4. Rewrite the output predictions as

$$\underset{\rightarrow}{y}(z) = \frac{b(z)\underset{\rightarrow}{\Delta u}(z) + p(z)}{A_+(z)A_-(z)} \tag{3.81}$$

from which it is clear that the output predictions are stable if and only if

$$b(z)\underset{\rightarrow}{\Delta u}(z) + p(z) = A_+ \phi(z) \tag{3.82}$$

where $\phi(z)$ is stable.

5. Eqn.(3.82) sets limitations on the possible choices of future control increments which are dependent on initial conditions given in $p(z)$.

(a) First find the minimal order solution to eqn.(3.82) assuming that both $\underset{\rightarrow}{\Delta u}(z)$, $\phi(z)$ are polynominals from

$$[\Gamma_b, \Gamma_{A_+}] \begin{bmatrix} \underset{\rightarrow}{\Delta u} \\ \phi \end{bmatrix} = p \quad \Rightarrow \quad \begin{bmatrix} \underset{\rightarrow}{\Delta u} \\ \phi \end{bmatrix} = \begin{bmatrix} P_1 \\ P_2 \end{bmatrix} p; \quad \begin{bmatrix} P_1 \\ P_2 \end{bmatrix} = [\Gamma_b, \Gamma_{A_+}]^{-1} \tag{3.83}$$

(b) The dimensions n_u, n_ϕ of the vectors $\underset{\rightarrow}{\Delta u}$, ϕ are determined from:

$$n_u + n_\phi = n_b + n_u = n_{A_+} + n_\phi \tag{3.84}$$

(c) As hinted before, matrices Γ_b, Γ_{A_+} have dimensions to give consistency and if necessary p is packed with zeros.

(d) The d.o.f. in the choices for $\underset{\rightarrow}{\Delta u}(z)$, $\phi(z)$ are as follows:

$$\underset{\rightarrow}{\Delta u} = \Gamma_{A_+} \underset{\rightarrow}{c}; \quad \phi = \Gamma_b \underset{\rightarrow}{c} \tag{3.85}$$

This is because these choices give $b(z)\underset{\rightarrow}{\Delta u}(z) = A_+ \phi(z)$; $\underset{\rightarrow}{c}$ can be any stable sequence (FIR or IIR).

(e) Equations (3.83, 3.85) imply that the overall solutions are given as

$$\underset{\rightarrow}{\Delta u} = P_1 p + \Gamma_{A_+} \underset{\rightarrow}{c}; \quad \phi = P_2 p + \Gamma_b \underset{\rightarrow}{c} \tag{3.86}$$

Summary: The choice of future control trajectory

$$\Delta \underrightarrow{u} = P_1 p + \Gamma_{A+} \underrightarrow{c} \tag{3.87}$$

guarantees the stability of the output *prediction* which is given from

$$\underrightarrow{y}(z) = \frac{[1, z^{-1}, \ldots](P_2 p + \Gamma_b \underrightarrow{c})}{A_-(z)} \tag{3.88}$$

Remark 3.5 *Care has been taken in this chapter to show that all the transfer function based prediction equations can be expressed in the form of (3.1) where $H = C_A^{-1} C_b$. Hence the observations above are applicable to them (including (3.75)) by appropriate change of p. For the interested reader, a specific application of the predictions of (3.75) using the structure in the problem appears in [133].*

3.7.2.2 Pole cancellation for the state-space case

In the state-space case, one needs to do an eigenvalue/vector decomposition to identify the unstable manifold. Hence define

$$A = [W_+, W_-] \begin{bmatrix} \Lambda_+ & 0 \\ 0 & \Lambda_- \end{bmatrix} \begin{bmatrix} V_+^T \\ V_-^T \end{bmatrix} \tag{3.89}$$

Now one can decouple the stable and unstable predictions (in the free mode, that is with $\Delta \underrightarrow{u} = 0$) as

$$x_{k+n} = \underbrace{W_+ \Lambda_+^n V_+^T x_0}_{\text{divergent}} + \underbrace{W_- V_-^T [A^{n-1} B, \ldots, B] \Delta \underrightarrow{u}}_{\text{convergent}} \tag{3.90}$$

From this is is clear that there is a need to parameterise the future control trajectory so that the part of $\underrightarrow{x}$ in the unstable manifold goes to zero, at some point beyond the control horizon (number of control moves in the prediction). Then there can be no divergent component in the predictions. Hence $\Delta \underrightarrow{u}$ should be chosen subject to the condition that:

$$V_+^T x_{k+n_u} = 0 \quad \Rightarrow \quad V_+^T (A^{n_u} x + [A^{n_u-1} B, \ldots, B] \Delta \underrightarrow{u}) = 0 \tag{3.91}$$

Remark 3.6 *The weakness of condition (3.91) is that it contains terms A^n which are divergent and hence should not be used for large n. However, neat ways around this do not give much additional insight for this book given the developments in Section 3.7.2.1, Chapter 7 (and [113]) and hence are omitted.*

Remark 3.7 *Numerical comparisons that illustrate the potential benefits of using pseudo pole cancellation in the prediction stage are given in Chapter 7.*

Summary: If you have an unstable closed-loop process, it maybe unwise to use open-loop predictions. Instead use some of the d.o.f. in the future control trajectory to prestabilise the predictions. This ensures a better conditionned problem.

3.8   Pseudo closed-loop prediction

This is an alternative to prestabilisation by pole cancellation which gives equivalent benefits in numerical conditioning, but has additional advantages with respect to optimality and stability analysis. For now, this book will not give any further details and the reader is referred to Chapter 7. The reader is left with the summary that one first prestabilises the process and then uses the inputs to the prestabilised loop as the d.o.f. in the predictions.

4

Predictive control – the basic algorithm

The purpose of this chapter is to give a brief summary of a standard predictive control algorithm. Most emphasis will be placed on generalised predictive control (GPC [21]), as the majority of standard algorithms are very similar in principle, the minor differences are mainly due to modelling/prediction assumptions. Brief sections will illustrate the changes required for other variants such as the popular industrial choice dynamic matrix control (DMC).

4.1 Summary of main results

For those readers who are interested in the results and do not want to read through the detail, they are given now. The details of how to compute the controller parameters follow in this chapter.

1. In the absence of constraints, a GPC control law reduces to a fixed linear feedback.

   (a) For transfer function based predictions, the control law takes the form:

   $$D_k(z)\Delta \mathbf{u} = P_r(z)\mathbf{r} - N_k(z)\mathbf{y} \qquad (4.1)$$

   (b) For state-space based predictions, the control law takes the form:

   $$\mathbf{u} - \mathbf{u}_{ss} = -K(\mathbf{x} - \mathbf{x}_{ss}) \qquad (4.2)$$

2. In the presence of constraints the optimum predicted control trajectory is defined through the on-line solution of a quadratic programming problem which takes the form:

   $$\min_{\underset{\rightarrow}{\Delta \mathbf{u}}} \ \Delta \mathbf{u}_{\rightarrow}^T S \Delta \mathbf{u}_{\rightarrow} + \Delta \mathbf{u}_{\rightarrow}^T \mathbf{p} \ \text{ s.t. } \ C \Delta \mathbf{u}_{\rightarrow} - \mathbf{d} \le 0 \qquad (4.3)$$

   where S is positive definite and $\mathbf{p}$, $\mathbf{d}$ are time varying (dependent on the current state).

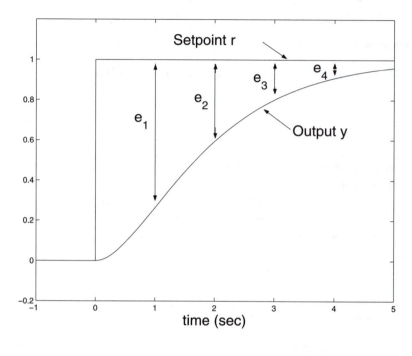

FIGURE 4.1
Tracking errors.

4.2 GPC algorithm – the main components

This section gives an overview of the main components in a predictive control law.
Use of these components to form the parameters of the control law then follows in
the later sections.

4.2.1 Performance index and optimisation

The control law is determined from the minimisation of a 2-norm measure of predicted performance. A typical performance index (or cost function) is

$$
\begin{aligned}
J &= \sum_{i=n_w}^{n_y} \|\mathbf{r}_{k+i} - \mathbf{y}_{k+i}\|_2^2 + \lambda \sum_{i=0}^{n_u-1} \|\Delta \mathbf{u}_{k+i}\|_2^2 \\
&= \sum_{i=n_w}^{n_y} \|\mathbf{e}_{k+i}\|_2^2 + \lambda \sum_{i=0}^{n_u-1} \|\Delta \mathbf{u}_{k+i}\|_2^2
\end{aligned}
\tag{4.4}
$$

That is:

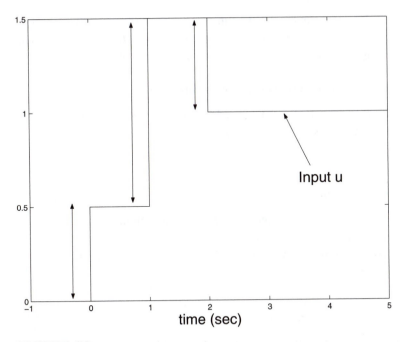

FIGURE 4.2

Input and input increments.

- Sum the squares of the predicted tracking errors from an initial horizon[*] n_w to an output horizon n_y.

- Sum the squares of the control changes over the horizon n_u.

It is assumed that control increments are zero beyond the control horizon, that is

$$\Delta \mathbf{u}_{k+i|k} = 0, \ \ i \geq n_u \tag{4.5}$$

4.2.1.1 Visualisation of terms in the performance index

The terms in the performance index are easy to visualise in graphical form. Assume a sampling instant of 1 sec; then:

- Figure 4.1 illustrates the first four tracking errors $\mathbf{e}_i$.

- Figure 4.2 illustrates the control moves for $n_u = 3$ (denoted by the double-sided arrows).

---

[*]It is common to assume $n_w = 1$ and this is the default assumed hereafter.

The aim of the performance index is to minimise a weighted sum of squares of these terms. Clearly the minimum performance index, that is $J = 0$, is consistent with offset free tracking, as it implies that both $\underrightarrow{e} = 0$ and the control is unchanging.

4.2.1.2   Compact representation of the performance index

The cost function is often better represented in more compact notation (e.g. see equations 1.1) using vectors and matrices. Hence J from (4.4) can be written as

$$J = \|\underrightarrow{r} - \underrightarrow{y}\|_2^2 + \lambda\|\Delta\underrightarrow{u}\|_2^2 = \|\underrightarrow{e}\|_2^2 + \lambda\|\Delta\underrightarrow{u}\|_2^2 \qquad (4.6)$$

4.2.1.3   The optimisation

The degrees of freedom (d.o.f.) which can be used to minimise J are the future control moves, typically the first n_u control moves. Hence the on-line control law is determined from the minimisation of the cost J w.r.t the n_u future control moves, that is $\Delta\underrightarrow{u}$. This minimisation is denoted as

$$\min_{\Delta\underrightarrow{u}} J = \|\underrightarrow{e}\|_2^2 + \lambda\|\Delta\underrightarrow{u}\|_2^2 \qquad (4.7)$$

Of the optimising $\Delta\underrightarrow{u}$, only the first element, that is Δu_k, is implemented as the optimisation is repeated (updated) at each sampling instant.

Summary: The cost to be minimised at each sampling instant usually takes the form of (4.6).

$$\min_{\Delta\underrightarrow{u}} J = \|\underrightarrow{e}\|_2^2 + \lambda\|\Delta\underrightarrow{u}\|_2^2 \qquad (4.8)$$

4.2.2   Restrictions on the predicted future control trajectory

It is important to note the restriction (4.5) placed on the class of future input trajectories. This has significant repercussions on the structure of the prediction equations summarised in (3.1). It is noted that the matrix H multiplies upon $\Delta\underrightarrow{u}$. Also for convenience the previous chapter allowed that H was square and that it multiplied on the entire future sequence of control moves.

Hence, the prediction equations assume that

$$H\Delta\underrightarrow{u} = \begin{bmatrix} h_0 & 0 & 0 & \cdots & 0 & \cdots & 0 \\ h_1 & h_0 & 0 & \cdots & 0 & \cdots & 0 \\ \vdots & \vdots & \vdots & \vdots & \vdots & \vdots & \vdots \\ h_{n_y-1} & h_{n_y-2} & h_{n_y-3} & \cdots & h_{n_y-n_u} & \cdots & h_0 \end{bmatrix} \begin{bmatrix} \Delta u_{k|k} \\ \Delta u_{k+1|k} \\ \vdots \\ \Delta u_{k+n_u-1|k} \\ \vdots \\ \Delta u_{k+n_y-1|k} \end{bmatrix} \qquad (4.9)$$

However, the restriction (4.5), that is $\Delta \mathbf{u}_{k+n_u+i|k} = 0, i \geq 0$, implies that

$$H\underset{\rightarrow}{\Delta \mathbf{u}} = \underbrace{\begin{bmatrix} h_0 & 0 & 0 & \cdots & 0 \\ h_1 & h_0 & 0 & \cdots & 0 \\ \vdots & \vdots & \vdots & \vdots & \vdots \\ h_{n_y-1} & h_{n_y-2} & h_{n_y-3} & \cdots & h_{n_y-n_u} \end{bmatrix}}_{H_{th}} \begin{bmatrix} \Delta \mathbf{u}_{k|k} \\ \Delta \mathbf{u}_{k+1|k} \\ \vdots \\ \Delta \mathbf{u}_{k+n_u-1|k} \end{bmatrix} \quad (4.10)$$

where H_{th} constitutes only the first n_u columns of H. Hereafter, for convenience, wherever the reader sees the prediction equation $H\underset{\rightarrow}{\Delta \mathbf{u}}$, it can be assumed that H is tall and thin, that is $H = H_{th}$.

Summary: In the term $H\underset{\rightarrow}{\Delta \mathbf{u}}$ of the prediction equations (e.g. (3.1)) it is assumed that H is block dimension $n_y \times n_u$ and the vector $\underset{\rightarrow}{\Delta \mathbf{u}}$ comprises block n_u terms.

4.2.3 The receding horizon concept

The terminology *receding horizon* is often applied to predictive control because one can imagine an MPC law as using a receding horizon, that is a horizon that is constantly moving away, although at the same speed at which you are moving. For instance

1. At sample k, the control law optimises predicted performance over the time span

$$k+1 \leq t \leq k+n_y$$

2. At sample $k+1$, the control law optimises performance over the time span

$$k+2 \leq t \leq k+n_y+1$$

Clearly the time span over which the optimisation takes place is always moving (receding). At the current sample one takes into account points that previously were beyond the time span. This is illustrated in Table 4.1 where clearly the start point and end point of the costing horizon recede.

4.2.4 Constraints

One of the major selling points of MPC is its ability to take systematic account of constraints, as these can easily be incorporated into the optimisation of (4.8). In many systems it is common to have constraints such as upper and lower limits on the input, e.g.

$$\underline{u}_i \leq u_i \leq \bar{u}_i \quad (4.11)$$

TABLE 4.1

Illustration of receding horizon

Sampling instant	Horizon window								
	0	1	2	3	4	5	6	7	8
0									
1									
2									
3									
$\vdots$									$\vdots$

On the input rates

$$\underline{\Delta u_i} \leq \Delta u_i \leq \overline{\Delta u_i} \tag{4.12}$$

One can also have constraints on outputs and states, e.g.

$$\underline{x_i} \leq x_i \leq \overline{x_i}; \quad \underline{y_i} \leq y_i \leq \overline{y_i} \tag{4.13}$$

Some constraints may even be a mixture of several outputs, for instance

$$y_1 + y_2 \leq \overline{y_{1,2}} \tag{4.14}$$

In this book we will take it for granted that optimisation (4.8) is in practice implemented as

$$\min_{\underline{\Delta \mathbf{u}}} J \quad \text{s.t. constraints} \tag{4.15}$$

where the constraints will depend on the process, but in general are assumed to be linear in the d.o.f. $\underrightarrow{\Delta \mathbf{u}}$. Hence (4.15) is a quadratic program, as the performance index is quadratic in the d.o.f. and the constraints are linear in the d.o.f. More details will be given later in this chapter.

Summary: The standard MPC optimisation (4.15) is a quadratic programming problem. It is accepted that this is a tractable problem in general, exceptions being processes with fast sampling rates or using large numbers of d.o.f.

4.2.5 Multivariable systems

The performance index J is written in such a way that it automatically allows for multivariable systems. The advantage of this is that MPC is a systematic design tool for handling interaction and stabilising multivariable systems. Very few alternatives are as systematic. In simple terms, MPC gives the best predicted performance by way of the objective set J for the given model. Implicitly interaction is dealt with in that high interaction would give high values of J and not be optimal. Likewise, predictions giving poor performance would not be optimal for sensibly chosen horizons.

In general there is always interaction in multivariable systems and moreover one may find that one loop is faster or more tightly controlled than another. These effects could be accentuated when the cost J is poorly selected. MPC algorithms have a facility to counter these undesirable effects, that is augmenting J with more specific weighting matrices (the original weighting λ in (4.6) is only scalar). For instance, rewrite J as

$$J = \sum_{i=n_w}^{n_y} \|W_y(\mathbf{r}_{k+i} - \mathbf{y}_{k+i})\|_2^2 + \lambda \sum_{i=0}^{n_u-1} \|W_u(\Delta\mathbf{u}_{k+i})\|_2^2 \qquad (4.16)$$

where matrix weights W_y, W_u are positive definite and diagonal. These weights can even vary (usually increase) with the horizon i though such a complication would rarely be justified, as the increase in the number of design parameters to select often gives negligible benefits. *Often* the sensitivity of the resulting controller parameters to the weights can be quite small so one may need to change the weights by an order of magnitude to see a significant effect.

The design/tuning of the diagonal elements of W_y, W_u to get a balance in the tracking performance of each loop and input activity of each loop, although straightforward in principle, can be tedious. In order to give a good start point it may be wise to first normalise all the signals in J so that the movement nought to one is equivalent in each loop and then initialise with $W_y = I$, $W_u = I$.

Note that the definition of *best* trade-off between response times in each loop and the size of the interactions is somewhat arbitrary unless one can produce a mathematical measure. If such a measure exists then MPC may well be able to use it directly but that topic is outside the scope of this book.

> **Summary:** MPC incorporates weights that allow a systematic design procedure for handling interaction in multivariable systems.

4.2.6 The use of input increments and obtaining integral action

The way that MPC incorporates integral action may seem a little obscure, so this section attempts to give a simple explanation. To get offset free tracking:

1. In steady state, the minimum of J must be consistent with zero tracking errors.

2. If the plant is in steady state with zero tracking errors, the predicted control move to maintain zero tracking errors must be zero. That is, the predictions must be unbiased.

The reader will note that there are two issues implicit in the above: (i) using a well-posed performance index and (ii) having unbiased predictions in steady state. It is easy to demonstrate why either of the above assumptions will cause offset in the closed-loop.

If the performance index is not set up so that the minimum (in steady state) corresponds to zero tracking error, then the converse must occur; that is, the *optimum* control will necessarily cause offset. One will note that J of (4.4) is well posed and this comes about by the use of $\Delta \underset{\rightarrow}{\mathbf{u}}$ rather than $\underset{\rightarrow}{\mathbf{u}}$. Hence, in steady-state, the minimum, $J = 0$, is given by $\underset{\rightarrow}{\mathbf{r}} = \underset{\rightarrow}{\mathbf{y}}$ and $\Delta \underset{\rightarrow}{\mathbf{u}} = 0$, that is, no predicted offset and no predicted change in the input.

The second issue is the need for unbiased predictions and this was discussed in the previous chapter. Assume that one is in steady state and $\mathbf{r} = \mathbf{y}$ and $\Delta \underset{\leftarrow}{\mathbf{u}} = 0$. One then wants that with $\Delta \underset{\rightarrow}{\mathbf{u}} = 0$ the prediction model should give $\underset{\rightarrow}{\mathbf{r}} = \underset{\rightarrow}{\mathbf{y}}$, regardless of any differences between the model and the process due to uncertainty and disturbances. This bias free prediction can be achieved by using the incremental model (see earlier chapter), as this is defined only to predict changes about the current values and does not use any information about the current absolute value of the input. Hence it is not susceptible to mismatch between model and process gains.

Summary:

1. If one uses tracking errors and *control increments* in the performance index, it will be well posed for ensuring offset free tracking in the control law.

2. Incremental models, those containing an internal disturbance model, give unbiased predictions in the steady state. This is essential for ensuring effective integral action in an MPC control law.

4.2.7   Eliminating tracking offset while weighting the inputs

One might ask: would using a performance index which weighted the inputs rather than the input increments conflict with the desire for integral action? The simple answer is that it would. Hence a performance index

$$J = \sum_{i=n_w}^{n_y} \|\mathbf{r}_{k+i} - \mathbf{y}_{k+i}\|_2^2 + \lambda \sum_{i=0}^{n_u-1} \|\mathbf{u}_{k+i}\|_2^2 \qquad (4.17)$$

would not be well posed in general as $\mathbf{r}_{k+i} = \mathbf{y}_{k+i}$ and $\mathbf{u}_{t+i} = 0$ will be inconsistent most of the time and hence minimising (4.17) cannot give offset free tracking [†]. The best minimum in the steady state would form a comprismise between the norms of $\mathbf{r} - \mathbf{y}$ and $\mathbf{u}$ and hence $\mathbf{r} - \mathbf{y} \neq 0$.

[†]An exception is if the process already has integral dynamics so that $\mathbf{u}_{+i} = 0$ in steady state for any setpoint and disturbance.

If one wants to avoid the use of input increments, then one alternative is to include weights on the distance of the inputs from their steady-state values $\mathbf{u}_{ss}$, e.g.

$$J = \sum_{i=n_w}^{n_y} \|\mathbf{r}_{k+i} - \mathbf{y}_{k+i}\|_2^2 + \lambda \sum_{i=0}^{n_u-1} \|\mathbf{u}_{k+i} - \mathbf{u}_{ss}\|_2^2 \qquad (4.18)$$

as then $\mathbf{r}_{k+i} = \mathbf{y}_{k+i}$ and $\mathbf{u}_{k+i} = \mathbf{u}_{ss}$ are consistent and the minimum $J = 0$ occurs with no offset. Of course one then needs a model that gives unbiased predictions for the pair $\mathbf{y}_{t+i}$, $\mathbf{u}_{ss}$. Models for giving such unbiased estimates of $\mathbf{u}_{ss}$ were discussed in Section 2.3.4.3 on state-space models.

Summary: The use of both performance indices (4.4) and (4.18) allow offset free control given unbiased predictions.

4.2.8 Links to optimal control

It is recognised that predictive control falls into the class of optimal control laws. However, whereas a true optimal control uses $n_y = n_u = \infty$ in the performance index, MPC algorithms tend not to make such assumptions (see Chapter 6 for exceptions). Hence, in practice MPC is suboptimal. However, if $n_y - n_u$ is greater than the open-loop settling time and $n_u \geq 5^{\ddagger}$, the differences are usually slight and unimportant in practice. The link to optimal control can be used to give a priori stability guarantees and this is discussed later.

Summary: For large $n_y - n_u$ and large n_u, GPC gives near identical control to an optimal control law with the same weights. For small $n_y - n_u$ and small n_u the resulting control law maybe severely suboptimal.

4.3 GPC algorithm formulation for transfer function models

This section is written assuming an MFD model and hence the SISO case is automatically included. The constraint free case only is considered next. First an outline is given of how to compute the GPC control law. For simplicity of notation the sample time subscript $(.)_k$ is omitted except where it is not implicit.

4.3.1 Steps to form a GPC control law

Step 1. Define the vector formulation of the cost eqn.(4.6)

$$J = \|\underset{\rightarrow}{\mathbf{r}} - \underset{\rightarrow}{\mathbf{y}}\|_2^2 + \lambda \|\Delta \underset{\rightarrow}{\mathbf{u}}\|_2^2 \qquad (4.19)$$

---

‡This is a rather arbitrary value and should be treated as such.

Step 2. Substitute $\underset{\rightarrow}{\mathbf{y}}$ from eqn.(3.25) into (4.19)

$$J = \| \underset{\rightarrow}{\mathbf{r}} - H\underset{\rightarrow}{\Delta\mathbf{u}} - P\underset{\leftarrow}{\Delta\mathbf{u}} - Q\underset{\leftarrow}{\mathbf{y}} \|_2^2 + \lambda \| \underset{\rightarrow}{\Delta\mathbf{u}} \|_2^2 \qquad (4.20)$$

Note (see Section 4.2.2) that H is tall and thin to take account of the fact that $\Delta\mathbf{u}_{k+i} = 0$, $i \geq n_u$.

Step 3. State the required minimisation after expanding (4.20):

$$\min_{\underset{\rightarrow}{\Delta\mathbf{u}}} \; J = \underset{\rightarrow}{\Delta\mathbf{u}}^T (H^T H + \lambda I)\underset{\rightarrow}{\Delta\mathbf{u}} + 2\underset{\rightarrow}{\Delta\mathbf{u}}^T H^T [P\underset{\leftarrow}{\Delta\mathbf{u}} + Q\underset{\leftarrow}{\mathbf{y}} - \underset{\rightarrow}{\mathbf{r}}] + k \qquad (4.21)$$

$k = \| \underset{\rightarrow}{\mathbf{r}} - P\underset{\leftarrow}{\Delta\mathbf{u}} - Q\underset{\leftarrow}{\mathbf{y}} \|_2^2$ contains terms that do not depend upon $\underset{\rightarrow}{\Delta\mathbf{u}}$ and hence can be ignored.

Step 4. Note that the performance index is quadratic (and always positive) and hence has a unique minimum which therefore can be located by setting the first derivative to zero:

$$\frac{dJ}{d\underset{\rightarrow}{\Delta\mathbf{u}}} = 2(H^T H + \lambda I)\underset{\rightarrow}{\Delta\mathbf{u}} + 2H^T [P\underset{\leftarrow}{\Delta\mathbf{u}} + Q\underset{\leftarrow}{\mathbf{y}} - \underset{\rightarrow}{\mathbf{r}}] \qquad (4.22)$$

$$\frac{dJ}{d\underset{\rightarrow}{\Delta\mathbf{u}}} = 0 \;\Rightarrow\; \underset{\rightarrow}{\Delta\mathbf{u}} = (H^T H + \lambda I)^{-1} H^T [\underset{\rightarrow}{\mathbf{r}} - P\underset{\leftarrow}{\mathbf{y}} - Q\underset{\leftarrow}{\Delta\mathbf{u}}] \qquad (4.23)$$

Step 5. The GPC control law is defined by the first element of $\underset{\rightarrow}{\Delta\mathbf{u}}$, i.e. $\Delta\mathbf{u}_k = \mathbf{e}_1^T \underset{\rightarrow}{\Delta\mathbf{u}}$, $\mathbf{e}_1^T = [I, 0, 0, ..., 0]$.

$$\Delta\mathbf{u}_k = \mathbf{e}_1^T (H^T H + \lambda I)^{-1} H^T [\underset{\rightarrow}{\mathbf{r}} - P\underset{\leftarrow}{\mathbf{y}} - Q\underset{\leftarrow}{\Delta\mathbf{u}}] \qquad (4.24)$$

Summary: The computation (4.24) is recalculated at each sampling instant and therefore the control law is

$$\Delta\mathbf{u}_k = P_r \underset{\rightarrow}{\mathbf{r}} - N_k \underset{\leftarrow}{\mathbf{y}} - \check{D}_k \underset{\leftarrow}{\Delta\mathbf{u}} \qquad (4.25)$$

where

$$\begin{aligned} P_r &= \mathbf{e}_1^T (H^T H + \lambda I)^{-1} H^T \\ N_k &= \mathbf{e}_1^T (H^T H + \lambda I)^{-1} H^T P \\ \check{D}_k &= \mathbf{e}_1^T (H^T H + \lambda I)^{-1} H^T Q \end{aligned} \qquad (4.26)$$

4.3.2 Transfer function representation of the control law

For implementation purposes, the formulation given in (4.25) is ideal and no further manipulations are required. However, for the purposes of closed-loop stability analysis, it is easier first to represent (4.25) in transfer function form. Given

$$
\begin{aligned}
N_k &= [N_{ko}, N_{k1}, N_{k2}, \ldots, N_{kn}] \\
\check{D}_k &= [D_{ko}, D_{k1}, D_{k2}, \ldots, D_{km}] \\
P_r &= [P_{r1}, P_{r2}, P_{r3}, \ldots, P_{rn_y}]
\end{aligned}
\tag{4.27}
$$

define

$$
\begin{aligned}
N_k(z) &= N_{ko} + N_{k1}z^{-1} + N_{k2}z^{-2} + \cdots + N_{kn}z^{-n} \\
\check{D}_k(z) &= \check{D}_{ko} + \check{D}_{k1}z^{-1} + \check{D}_{k2}z^{-2} + \cdots + \check{D}_{km}z^{-m} \\
P_r(z) &= P_{r1}z + P_{r2}z^2 + P_{r3}z^3 + \cdots + P_{rn_y}z^{n_y} \\
D_k(z) &= 1 + z^{-1}\check{D}_k(z)
\end{aligned}
\tag{4.28}
$$

Then, noting the definitions (1.1, 3.21) of $\underleftarrow{\mathbf{u}}, \underleftarrow{\mathbf{y}}, \underrightarrow{\mathbf{r}}$, the control law can be implemented in the following fixed closed loop form.

$$
D_k(z)\Delta\mathbf{u}_k = P_r(z)\mathbf{r}_k - N_k(z)\mathbf{y}_k
\tag{4.29}
$$

It should be noted that $P_r(z)$ is an anticausal operator; that is, it uses future values of **r**. If these are unknown, simply subsitute the best available estimate, for instance the current value.

Remark 4.1 *We note [109] that the default choice for P_r given in GPC is often poor and better alternatives exist when advance knowledge is available. It is often as convenient to adopt a performance index*

$$
J = \|\underrightarrow{\mathbf{y}}\|_2^2 + \lambda\|\underrightarrow{\Delta\mathbf{u}}\|_2^2
\tag{4.30}
$$

and cater for tracking (i.e. nonzero r) by alternative means. For instance, if no advance knowledge of the set point is given, then we must have $P_r = N_k(1)$.

Summary: In the absence of constraints, GPC reduces to a known and fixed linear feedback law with feedforward.

$$
D_k(z)\Delta\mathbf{u}_k = P_r(z)\mathbf{r}_k - N_k(z)\mathbf{y}_k
\tag{4.31}
$$

4.3.3 Closed-loop transfer functions

In order to do a sensitivity or closed-loop pole analysis, one may want to find the closed-loop transferences. These are of course easy to derive given the control law (4.31). However, for completeness, this section outlines the transference from **r** to **y**. The argument $(.)(z)$ is omitted to improve readability.

Let the MFD model $D\mathbf{y} = N\mathbf{u}$ be written in two forms

$$
\hat{N}\hat{D}^{-1} = D^{-1}N \;\Rightarrow\;
\begin{cases}
\mathbf{y} = \hat{N}\hat{D}^{-1}\mathbf{u} \\
\mathbf{y} = D^{-1}N\mathbf{u}
\end{cases}
\tag{4.32}
$$

The equations in the loop are:

$$\underbrace{\mathbf{y} = \hat{N}\hat{D}^{-1}\mathbf{u};}_{\text{Model}} \quad \underbrace{D_k\Delta\mathbf{u} = P_r\mathbf{r} - N_k\mathbf{y};}_{\text{Controller}} \tag{4.33}$$

Therefore

$$[D_k\Delta]\mathbf{u} = P_r\mathbf{r} - N_k\hat{N}\hat{D}^{-1}\mathbf{u} \quad \Rightarrow \quad [D_k\Delta\hat{D} + N_k\hat{N}]\hat{D}^{-1}\mathbf{u} = P_r\mathbf{r} \tag{4.34}$$

Hence

$$\begin{aligned}\mathbf{u} &= \hat{D}[D_k(z)\Delta\hat{D} + N_k(z)\hat{N}]^{-1}P_r(z)\mathbf{r} \\ \mathbf{y} &= \hat{N}[D_k(z)\Delta\hat{D} + N_k(z)\hat{N}]^{-1}P_r(z)\mathbf{r}\end{aligned} \tag{4.35}$$

Clearly the closed-loop poles depend on $\det(D_k(z)\Delta\hat{D} + N_k(z)\hat{N})$.

Remark 4.2 *In general matrix multiplication is not commutative, i.e. $AB \neq BA$. Hence in the manipulations of this section, one must be careful to preserve the order of the operations.*

Summary: Given (4.31, 4.32), the closed-loop poles can be derived from

$$P_c(z) = D_k(z)\Delta\hat{D} + N_k(z)\hat{N} \tag{4.36}$$

4.3.4   GPC based on MFD models with a T-filter (GPCT)

The reader will recall that the inclusion of a T-filter modified the prediction equations, so that one would use predictions (3.33) in place of (3.21). Therefore there is a need to redo the algebra of Section 4.3.1 using the modified predictions. However, it is also clear that the only change in the prediction equations is as follows:

$$(P\underleftarrow{\mathbf{y}} + Q\Delta\underleftarrow{\mathbf{u}}) \quad \rightarrow \quad (\tilde{P}\underleftarrow{\tilde{\mathbf{y}}} + \tilde{Q}\Delta\underleftarrow{\tilde{\mathbf{u}}}) \tag{4.37}$$

Therefore one can simply substitute this change into (4.25) and derive the modified control law accordingly.

By analogy with (4.25, 4.26), the corresponding control law takes the form:

$$\begin{aligned}\Delta\mathbf{u}_k &= P_r\underrightarrow{\mathbf{r}} - \check{D}_k\Delta\underleftarrow{\tilde{\mathbf{u}}} - \tilde{N}_k\underleftarrow{\tilde{\mathbf{y}}} \\ \begin{cases} \check{D}_k = [I,0,0,\ldots][H^T H + \lambda I]^{-1}H^T\tilde{P} \\ \tilde{N}_k = [I,0,0,\ldots][H^T H + \lambda I]^{-1}H^T\tilde{Q} \end{cases}\end{aligned} \tag{4.38}$$

Substituting $\tilde{\mathbf{u}} = \mathbf{u}/T$, $\tilde{\mathbf{y}} = \mathbf{y}/T$ and rearranging into a more conventional form in terms of z-transforms (as in Section 4.3.2) and omitting the argument $.(z)$, gives:

$$[1 + \frac{\check{D}_k}{T}]\Delta\mathbf{u} = P_r\mathbf{r} - \frac{\tilde{N}_k}{T}\mathbf{y} \tag{4.39}$$

or more conveniently

$$\frac{\tilde{D}_k \Delta}{T} \mathbf{u} = P_r \mathbf{r} - \frac{\tilde{N}_k}{T} \mathbf{y}; \quad \tilde{D}_k = T + \check{D}_k \tag{4.40}$$

Note that the corresponding control parameters for GPC (D_k and N_k) and GPCT ($\tilde{D}_k$ and $\tilde{N}_k$ will be different, as $T \neq 1$ implies $\tilde{P} \neq P$, $\tilde{Q} \neq Q$ — see equation (3.33). Nevertheless, in the absence of model uncertainty and disturbances, control laws (4.31, 4.40) give identical tracking performance.

One can form the closed-loop poles arising from combining model (4.32) with controller (4.40) by analogy to (4.35). Clearly by inspection:

$$\begin{aligned} \mathbf{u} &= \hat{D} T [\tilde{D}_k \Delta \hat{D} + \tilde{N}_k \hat{N}]^{-1} P_r \mathbf{r} \\ \mathbf{y} &= \hat{N} T [\tilde{D}_k \Delta \hat{D} + \tilde{N}_k \hat{N}]^{-1} P_r(z) \mathbf{r} \end{aligned} \tag{4.41}$$

where one notes that the only differences are the insertion of T and the use of $\tilde{D}_k$ and $\tilde{N}_k$ in place of D_k and N_k. The closed-loop poles are given by

$$\tilde{P}_c = \tilde{D}_k(z) \Delta \hat{D} + \tilde{N}_k(z) \hat{N} \tag{4.42}$$

Remark 4.3 *There is a relationship between $\tilde{P}_c$ and P_c of (4.36). In the nominal case (4.35) and (4.41) must be identical because an identical performance index was minimised with identical d.o.f. Hence the roots of $\tilde{P}_c$ include the roots of T.*

Remark 4.4 *Performance of (4.35) and (4.41) differs for the uncertain case, that is when $\hat{N}$, $\hat{D}$ do not correspond to those used for forming the prediction equations (3.21) and (3.33) or when there are disturbances. This is because in such a case GPC and GPCT will be using different predictions and hence the optimisations give differing answers.*

Summary: The z-transform representation of the control laws for GPC, GPCT are summarised in the table below.

	Control laws	**Closed-loop pole polynomial**
GPC	$D_k \Delta \mathbf{u} = P_r \mathbf{r} - N_k \mathbf{y}$	$D_k(z) \Delta \hat{D} + N_k(z) \hat{N}$
GPCT	$\dfrac{\tilde{D}_k}{T} \Delta \mathbf{u} = P_r \mathbf{r} - \dfrac{\tilde{N}_k}{T} \mathbf{y}$	$\tilde{D}_k(z) \Delta \hat{D} + \tilde{N}_k(z) \hat{N}$

4.4 Predictive control with a state-space model

This section gives two alternative means of setting up a state-space based predictive control law based on either state augmentation and performance index (4.6) or no state augmentation and performance index (4.18). It is assumed that having gained some expertise, readers will be able to modify what is included here to meet their own aims.

4.4.1   Simple state augmentation

In the transfer function case use is made of an incremental model and this automatically is written in terms of input increments which can be substituted into a cost function of the type (4.4). In the state-space case it is more usual to use state augmentation of the model, in essence making $\mathbf{u}_k$ an additional internal state and $\Delta\mathbf{u}$ the input d.o.f. One can achieve this by rewriting the state-space equation. Several means of doing this have been proposed, for example one can use the augmented state-space model

$$\begin{bmatrix} \mathbf{x}_{k+1} \\ \mathbf{u}_k \end{bmatrix} = \underbrace{\begin{bmatrix} A & B \\ 0 & I \end{bmatrix}}_{\hat{A}} \underbrace{\begin{bmatrix} \mathbf{x}_k \\ \mathbf{u}_{k-1} \end{bmatrix}}_{\hat{\mathbf{x}}} + \underbrace{\begin{bmatrix} B \\ I \end{bmatrix}}_{\hat{B}} \Delta\mathbf{u}_k \qquad (4.43)$$

$$\mathbf{y}_k = \underbrace{[C \ \ D]}_{\hat{C}} \begin{bmatrix} \mathbf{x}_k \\ \mathbf{u}_{k-1} \end{bmatrix} + D\Delta\mathbf{u}_k + \mathbf{d}_k$$

To ensure no bias in steady-state predictions, this model should satisfy prediction consistency conditions (Section 4.2.6); that is

$$\mathbf{y} = \mathbf{r} \text{ and } \Delta\mathbf{u} = 0 \qquad (4.44)$$

As the disturbance $\mathbf{d}_k$ varies, the implied steady-state values of $\mathbf{u}$ and $\mathbf{x}$ in the model (4.43) can move to ensure (4.44) holds.

4.4.1.1   Computing the predictive control law

The GPC algorithm can now be implemented in a straightforward fashion using similar steps to those for the MFD model (Section 4.3.1).

Step 1. Find prediction equations for the augmented model (4.43) using (3.10) and substitute these into J of eqn.(4.6) to give

$$J = \| \underset{\rightarrow}{\mathbf{r}} - H\underset{\rightarrow}{\Delta\mathbf{u}} - P\hat{\mathbf{x}}_k - L\mathbf{d} \|_2^2 + \lambda \| \underset{\rightarrow}{\Delta\mathbf{u}} \|_2^2 \qquad (4.45)$$

where L has the usual definition (see Chapter 3).

Step 2. Perform the optimisation of miminising J w.r.t. $\underset{\rightarrow}{\Delta\mathbf{u}}$, hence

$$\frac{dJ}{d\underset{\rightarrow}{\Delta\mathbf{u}}} = 0 \ \Rightarrow \ (H^T H + \lambda I)\underset{\rightarrow}{\Delta\mathbf{u}} = [H^T \underset{\rightarrow}{\mathbf{r}} - H^T P\hat{\mathbf{x}}_k - H^T L\mathbf{d}] \qquad (4.46)$$

Step 3. Solve for the first element of $\Delta \underset{\rightarrow}{\mathbf{u}}$

$$\Delta \mathbf{u}_k = \mathbf{e}_1^T (H^T H + \lambda I)^{-1} H^T [\underset{\rightarrow}{\mathbf{r}} - [P,L] \begin{bmatrix} \hat{\mathbf{x}}_k \\ \mathbf{d} \end{bmatrix}]$$
$$= P_r \underset{\rightarrow}{\mathbf{r}} - \hat{K} \begin{bmatrix} \hat{\mathbf{x}}_k \\ \mathbf{d} \end{bmatrix} \tag{4.47}$$

where $\hat{K} = \mathbf{e}_1^T (H^T H + \lambda I)^{-1} H^T [P,L], \; P_r = \mathbf{e}_1^T (H^T H + \lambda I)^{-1} H^T$.

Hence predictive control reduces to a state feedback.

Remark 4.5 *It is easy to see that P_r from step 3 above is identical to that of (4.26).*

Summary: Predictive control based on a state-space model takes the form of a state feedback (4.47) of the augmented state plus some feedforward.

4.4.1.2 Closed-loop equations and integral action

The nominal closed loop can be derived by combining (4.43) and (4.47). Hence

$$\hat{\mathbf{x}}_{k+1} = \hat{A}\hat{\mathbf{x}}_k - \hat{B}\hat{K} \begin{bmatrix} \hat{\mathbf{x}}_k \\ \mathbf{d} \end{bmatrix} + \hat{B}P_r \underset{\rightarrow}{\mathbf{r}}; \quad \mathbf{y}_k = \hat{C}\hat{\mathbf{x}}_k + D\Delta\mathbf{u}_k + \mathbf{d}_k \tag{4.48}$$

Hence the poles can be derived from the eigenvalues of $[\hat{A} - \hat{B}\hat{K}E]$, where E selects the relevant columns of $\hat{K}$. However, the reader may still be concerned at the apparent lack of attention given to the disturbance $\mathbf{d}$ in the predictions.

Offset free tracking in the steady state is assured by the consistency of (4.43) in the state estimator (model) and hence carries over to uncertain plant [§]. It is noted therefore that the state estimator must make an unbiased estimate for $\mathbf{d}$ and hence this introduces the integrator (pole at one) through the estimator. The estimator can be based on the original model (3.2) with state $\mathbf{x}$, as the part of the augmented state $\mathbf{u}_k$ is known exactly so does not need to be estimated.

[§]Although the real plant is subject to uncertainty, the consistency of the prediction model ensures integral action.

Summary:

1. For a state-space model the GPC control law takes the form

$$\Delta \mathbf{u}_k = P_r \underset{\rightarrow}{\mathbf{r}} - \hat{K} \begin{bmatrix} \hat{\mathbf{x}}_k \\ \mathbf{d} \end{bmatrix} \qquad (4.49)$$

   where $\hat{\mathbf{x}}$ is an augmented state and $\mathbf{d}$ is the disturbance estimate.

2. Integral action and offset free control are included by having an internal model of the disturbance (which must be estimated) and ensuring consistency of (4.44) so that the minimum achievable cost is $J = 0$.

4.4.2   State-space models without state augmentation

The state augmentation approach is analogous to the incremental model approach used in the transfer function models, though clearly there is some flexibility in exactly how the state is augmented. There are however alternatives, which may be considered more favourable. Consider the underlying requirements: (i) include a disturbance model; (ii) use a model giving offset free (unbiased) prediction and (iii) ensure the minimum of the cost function is consistent with zero tracking errors and a modified performance index, that is (4.18).

One can achieve these three requirements by making appropriate use of unbiased (e.g. Section 2.3.4) estimates $\mathbf{x}_{ss}$, $\mathbf{u}_{ss}$ [90] of the steady-state values for the state and input to give zero tracking errors.

- Form an algorithm to give estimates of $\mathbf{x}_{ss}$, $\mathbf{u}_{ss}$ which are consistent with zero tracking errors. The estimates are for the estimator and hence robust to model uncertainty.

- Use the cost function and optimisation

$$\min_{\underset{\rightarrow}{\mathbf{u}}} \; J = [\underset{\rightarrow}{\mathbf{x}} - \mathbf{x}_{ss}]^T Q [\underset{\rightarrow}{\mathbf{x}} - \mathbf{x}_{ss}] + [\underset{\rightarrow}{\mathbf{u}} - \mathbf{u}_{ss}]^T R [\underset{\rightarrow}{\mathbf{u}} - \mathbf{u}_{ss}] \qquad (4.50)$$

   which penalises deviations from the desired steady state. This differs from (4.6) in that the weighting of the inputs optimises the distance from steady state rather than rate of change (increment magnitude). It is clear that minimisation of this cost is consistent with zero tracking errors. It is commonly taken that Q comprises terms of the form $C^T C$ where $\mathbf{r} - \mathbf{y} = C(\mathbf{x}_{ss} - \mathbf{x})$.

4.4.2.1   Control law calculations

Lemma 4.1 *Let J be given as*

$$\min_{\underset{\rightarrow}{\mathbf{u}}}\ J = \underset{\rightarrow}{\mathbf{x}}^T Q \underset{\rightarrow}{\mathbf{x}} + \underset{\rightarrow}{\mathbf{u}} R \underset{\rightarrow}{\mathbf{u}} \tag{4.51}$$

The minimisation of J is known to give a state feedback control law of the form $\mathbf{u} = -K\mathbf{x}$. *Therefore the minimisation (4.50) must give a solution of the form*

$$\mathbf{u} = -K(\mathbf{x} - \mathbf{x}_{ss}) + \mathbf{u}_{ss} \tag{4.52}$$

Proof: This is obvious, as we have used a simple translation of the target which gives a simple and corresponding translation of the optimal control law. $\quad\square$

Theorem 4.1 *Minimisation (4.50) using predictions (3.10) gives the control law (4.52) with*

$$K = [H_x^T Q H_x + R]^{-1} H^T Q P_{xx} \tag{4.53}$$

Proof: Minimisation of (4.51) with prediction equations (3.8) reduces to:

$$\min_{\underset{\rightarrow}{\mathbf{u}}}\ J = [P_{xx}\mathbf{x} + H_x \underset{\rightarrow}{\mathbf{u}}]^T Q [P_{xx}\mathbf{x} + H_x \underset{\rightarrow}{\mathbf{u}}] + \underset{\rightarrow}{\mathbf{u}}^T R \underset{\rightarrow}{\mathbf{u}} \tag{4.54}$$

which gives an optimal control law

$$\underset{\rightarrow}{\mathbf{u}} = -[H_x^T Q H_x + R]^{-1} H_x^T Q P_{xx}\mathbf{x} = -K\mathbf{x} \tag{4.55}$$

Then using Lemma 4.1 gives the result. $\quad\square$

Note: State feedback (4.55) differs from (4.47) because it uses prediction equations (3.10) based on (3.2) instead of predictions (3.10) based on (4.43). That is, there is no state augmentation. Moreover the disturbance estimate is incorporated via $\mathbf{x}_{ss}, \mathbf{u}_{ss}$.

Summary: The predictive control law can be implemented as

$$\mathbf{u}_k - \mathbf{u}_{ss} = -\mathbf{e}_1^T [H_x^T Q H_x + R]^{-1} H_x^T Q P_{xx}(\mathbf{x} - \mathbf{x}_{ss}) \tag{4.56}$$

4.4.2.2   Estimating steady-state values for the state and input

This section gives a brief reminder of how to ensure unbiased estimates of the steady state values by using an appropriate disturbance model in the observer. If one can get an unbiased estimate of the disturbance, then the predictions will be unbiased, in the steady state.

1. Add a disturbance to the state-space model as follows:

$$\mathbf{x}_{k+1} = A\mathbf{x}_k + B\mathbf{u}_k; \quad \mathbf{d}_{k+1} = \mathbf{d}_k; \quad \mathbf{y}_k = C\mathbf{x}_k + D\mathbf{u}_k + \mathbf{d}_k \tag{4.57}$$

where **d** is an unknown signal representing disturbances and model mismatch.

2. Build an observer to estimate $\mathbf{x}, \mathbf{d}$. Let the observer take the form

$$\mathbf{z}_{k+1} = \tilde{A}\mathbf{z}_k + \tilde{B}\mathbf{u}_k - L(\tilde{C}\mathbf{z}_k + D\mathbf{u}_k - \mathbf{y}_k) \qquad (4.58)$$

where

$$\mathbf{z}_{k+1} \approx \begin{bmatrix} \mathbf{x}_{k+1} \\ \mathbf{d}_{k+1} \end{bmatrix}; \quad \tilde{A} = \begin{bmatrix} A & 0 \\ 0 & I \end{bmatrix}; \quad \tilde{B} = \begin{bmatrix} B \\ 0 \end{bmatrix}; \quad \tilde{C} = \begin{bmatrix} C & I \end{bmatrix} \qquad (4.59)$$

Clearly $\mathbf{z}$ contains the estimates of the true state $\mathbf{x}$ and disturbance $\mathbf{d}$. L is the observer gain to be designed either by Kalman methods or otherwise.

3. Given that $\mathbf{d}$, $\mathbf{r}$ are known, estimate the required steady-state values of $\mathbf{x}, \mathbf{u}$ to get desired output, i.e. $\mathbf{y} = \mathbf{r}$, from the estimator

$$\mathbf{y} = \mathbf{r} = C\mathbf{x}_{ss} + D\mathbf{u}_{ss} + \mathbf{d}; \quad \mathbf{x}_{ss} = A\mathbf{x}_{ss} + B\mathbf{u}_{ss} \qquad (4.60)$$

4. One can compute $\mathbf{x}_{ss}$, $\mathbf{u}_{ss}$ by solving (4.60) as simultaneous equations.

5. The control law is implemented as

$$\mathbf{u} = -K(\mathbf{x} - \mathbf{x}_{ss}) + \mathbf{u}_{ss} \qquad (4.61)$$

Theorem 4.2 *If one drives state estimates to $\mathbf{x}_{ss}$, $\mathbf{u}_{ss}$, then the output $\mathbf{y}$ must converge to $\mathbf{r}$, even if the model is uncertain.*

Proof: The observer update equation is given from

$$\mathbf{z}_{k+1} = \tilde{A}\mathbf{z}_k + \tilde{B}\mathbf{u}_k - L(\tilde{C}\mathbf{z}_k + D\mathbf{u}_k - \mathbf{y}_k) \qquad (4.62)$$

This uses $\mathbf{y}$ from the actual plant. Assume the process output is steady, then the state estimate $\mathbf{z}$ will settle if and only iff

$$\tilde{C}\mathbf{z}_k + D\mathbf{u}_k = \mathbf{y}_k; \quad \mathbf{u}_k = \mathbf{u}_{k-1} = \cdots \qquad (4.63)$$

This in turn implies that the observer output must equal to the process output. However, steady state also implies (from 4.60) that $\mathbf{x} = \mathbf{x}_{ss}$, $\mathbf{u} = \mathbf{u}_{ss}$ and therefore the values $\mathbf{x}_{ss}$, $\mathbf{u}_{ss}$ are consistent with an observer output equal to the set point $\mathbf{r}$. This in turn must imply that the process output is $\mathbf{r}$. $\qquad \square$

Remark 4.6 *One can also set up the disturbance model on the states or inputs if that is more appropriate see [90] or Section 2.3.4.*

Summary: Consistent steady-state values $\mathbf{x}_{ss}$, $\mathbf{u}_{ss}$ can be computed from (4.60). Use of these in conjunction with observer (4.58) and control law (4.61) gives offset free tracking in the steady state.

4.5   Formulation for finite impulse response models

This can be stated by inspection given the results earlier in this chapter, as a finite impulse response (FIR) model is the same as an MFD model with particular choices of D, N. Again remember the requirements for the cost function is that the minimum is consistent with zero tracking errors and the need for offset free prediction in the steady state. The former requirement is satisfied by a cost of the form (4.6) and offset free prediction for FIR models was discussed in the previous chapter (Section 3.5). The GPC algorithm for FIR models can now be summarised.

Step 1. Take the prediction equation of (3.61)

$$\underrightarrow{\mathbf{y}} = C_H \underrightarrow{\Delta \mathbf{u}} + M \underleftarrow{\Delta \mathbf{u}} + L \hat{\mathbf{y}} \tag{4.64}$$

and note that $C_H = H$ (where H is given in for instance eqn.(3.21) and Section 4.2.2).

Step 2. Subsitute this into the cost (4.6). Hence

$$J = [\underrightarrow{\mathbf{r}} - L\hat{\mathbf{y}} - H\underrightarrow{\Delta \mathbf{u}} - M\underleftarrow{\Delta \mathbf{u}}]^T [\underrightarrow{\mathbf{r}} - L\hat{\mathbf{y}} - H\underrightarrow{\Delta \mathbf{u}} - M\underleftarrow{\Delta \mathbf{u}}] + \lambda \underrightarrow{\Delta \mathbf{u}}^T \underrightarrow{\Delta \mathbf{u}} \tag{4.65}$$

Step 3. Minimise J w.r.t. $\underrightarrow{\Delta \mathbf{u}}$ to give

$$\underrightarrow{\Delta \mathbf{u}} = [H^T H + \lambda I]^{-1} H^T [\underrightarrow{\mathbf{r}} - L\hat{\mathbf{y}} - M\underleftarrow{\Delta \mathbf{u}}] \tag{4.66}$$

Step 4. Extract the first element $\Delta \mathbf{u}_k$

$$\begin{aligned} \Delta \mathbf{u}_k &= \mathbf{e}_1^T [H^T H + \lambda I]^{-1} H^T [\underrightarrow{\mathbf{r}} - L\hat{\mathbf{y}} - M\underleftarrow{\Delta \mathbf{u}}] \\ &= P_r \underrightarrow{\mathbf{r}} - \check{D}_k \underleftarrow{\Delta \mathbf{u}} - N_k \hat{\mathbf{y}} \end{aligned} \tag{4.67}$$

One notes that this has the same form as (4.25) with the only difference being the orders of $N_k, \check{D}_k$. That is less emphasis is placed on measured outputs (N_k has just one term) and more emphasis is placed on past inputs ($\check{D}_k$ has the same number of terms as the FIR model).

Remark 4.7 *The control law of (4.67) is more commonly known as DMC [23]. The only significant difference with GPC [21] is the model used.*

Summary: GPC for FIR models takes the same form as GPC for MFD models except that the compensators have different orders.

4.6   Formulation for independent models

As stated in the earlier chapters an independent model (IM) can take the form of any model we wish, transfer function, MFD, FIR, state-space, etc. Hence a GPC algorithm based on an IM will make use of the corresponding control law formulation for the given model type, e.g. (4.25, 4.52, 4.67,...) as appropriate. There is, however, one important difference which is obvious from consideration of the prediction equations given in the prediction chapter (Section 3.6).

Much like with an observer, an IM has states and outputs which are different from the process states and outputs and the offset between the process and the IM must be included to give unbiased predictions. Let $\check{\mathbf{y}}$ be the output of the IM and $\mathbf{y}$ be the measured process output.

4.6.1   IM is a transfer function or MFD

Take the prediction equation of (3.64)

$$\underrightarrow{\mathbf{y}} = H\Delta\underrightarrow{\mathbf{u}} + P\Delta\underleftarrow{\mathbf{u}} + Q\underleftarrow{\check{\mathbf{y}}} + L[\mathbf{y} - \check{\mathbf{y}}] \qquad (4.68)$$

One notes that this differs in form from (3.21) due to the additional term $L[\mathbf{y}_k - \check{\mathbf{y}}_k]$ to compensate for any bias caused by Q acting on $\underleftarrow{\check{\mathbf{y}}}$ as opposed to $\underleftarrow{\mathbf{y}}$. Substituting (4.68) into (4.6) and using the results of Section 4.3, one can state the the resulting control law by inspection as

$$\Delta\underrightarrow{\mathbf{u}} = [H^T H + \lambda I]H^T[\underrightarrow{\mathbf{r}} - P\Delta\underleftarrow{\mathbf{u}} - Q\underleftarrow{\check{\mathbf{y}}} - L\mathbf{y}_k + L\check{\mathbf{y}}_k] \qquad (4.69)$$

or in terms of z-transforms (using obvious substitutions)

$$\Delta\mathbf{u}_k = P_r\underrightarrow{\mathbf{r}} - \check{D}_k\Delta\underleftarrow{\mathbf{u}} - \check{N}_k\underleftarrow{\check{\mathbf{y}}} - M_k\mathbf{y} + M_k\check{\mathbf{y}} \qquad (4.70)$$

One can group common terms to give

$$D_k\Delta\mathbf{u} = P_r\underrightarrow{\mathbf{r}} - -N_k\underleftarrow{\check{\mathbf{y}}} - M_k\mathbf{y} \qquad (4.71)$$

where $D_k = [1 + z^{-1}\tilde{D}_k]$, $N_k = [\check{N}_k - M_k]$. Of course it is implicit in this control law that the IM model update is actually part of the control law calculation, so one must include the following IM equations in the compensator

$$D(z)\check{\mathbf{y}} = N(z)\mathbf{u}(z); \quad \mathbf{u}_k = \mathbf{u}_{k-1} + \Delta\mathbf{u}_k \qquad (4.72)$$

> **Summary:** For MFD models and an IM prediction, the GPC compensator takes the form of the following equations:
>
> $$D_k \Delta \mathbf{u} = P_r \underrightarrow{\mathbf{r}} - N_k \underleftarrow{\mathbf{\check{y}}} - M_k \mathbf{y}$$
> $$D(z)\mathbf{\check{y}} = N(z)\mathbf{u}(z) \tag{4.73}$$
> $$\mathbf{u}_k = \mathbf{u}_{k-1} + \Delta \mathbf{u}_k$$

4.6.2 Closed-loop poles in the IM case with an MFD model

One can compute the closed-loop poles by combining the compensator equations (4.73) with the model. First simplify the compensator into a single relationship by introducing the IM (4.72) into (4.71) and eliminate the IM output $\mathbf{\check{y}}$.

$$D_k(z)\Delta \mathbf{u} = P_r \mathbf{r} - N_k(z)[D^{-1}N]\mathbf{u} - M_k(z)\mathbf{y} \tag{4.74}$$

Collect common terms

$$D_i \mathbf{u} = P_r \mathbf{r} - M_k \mathbf{y}; \quad D_i = [D_k \Delta + N_k D^{-1} N] \tag{4.75}$$

Now compensator (4.75) is in a similar form to the control laws for GPC and GPCT (e.g. (4.31)) and thus we can use analogies to derive sensitivity and closed-loop poles. Hence, by inspection (say from (4.36)), the closed-loop poles for model (4.32) in conjunction with compensator (4.73) can be computed from

$$P_c = D_i \hat{D} + M_k \hat{N} = [D_k \Delta + N_k D^{-1} N] \hat{D} + M_k \hat{N}$$
$$= D_k \Delta \hat{D} + N_k \hat{N} + M_k \hat{N} \tag{4.76}$$

This can be simplified to

$$P_c = D_k \Delta \hat{D} + \underbrace{[N_k + M_k]}_{N_{im}} \hat{N} \tag{4.77}$$

which is unsurprisingly identical in form to (4.36).

> **Summary:** The nominal closed-loop poles for compensation based on an IM model are the same as with a more conventional use of the MFD model and the implied compensator can be reduced to a similar form.

4.6.3 IM is a state-space model

For use of a cost function (4.6), the appropriate prediction equations are modified to

$$\underrightarrow{\mathbf{y}} = P\hat{\mathbf{x}} + H\Delta \underrightarrow{\mathbf{u}} + L[\mathbf{y}_k - \mathbf{\check{y}}_k] \tag{4.78}$$

where augmented state $\hat{\mathbf{x}}$ includes the current value of $\mathbf{u}$. There is no need for an observer to estimate $\mathbf{x}$, as the IM model states are known and there is no need to estimate $\mathbf{d}$ or $\mathbf{x}_{ss}, \mathbf{u}_{ss}$, as this role (disturbance modelling) is taken by the bias correction term $L[\mathbf{y}_k - \check{\mathbf{y}}_k]$. The control law therefore takes the form

$$\Delta \mathbf{u}_k = P_r \underset{\rightarrow}{\mathbf{r}} - \hat{K}\mathbf{x} - P_r(1)[\mathbf{y}_k - \check{\mathbf{y}}_k] \tag{4.79}$$

Remark 4.8 *One can form similar relationships for use with cost function (4.50).*

4.7 General comments on stability analysis of GPC

The closed-loop poles for *unconstrained* GPC can be computed for both the state-space and transfer function prediction models (e.g. 4.36, 4.48). This is because the control law reduces to a fixed linear feedback law. However, there are few generic a priori stability results with GPC; that is to say, for an arbitrary set of n_y, n_u, λ the resulting control law may be destabilising or give poor performance. The next chapter will discuss means of avoiding this apparent weakness and in this chapter we have simply stated how the closed-loop poles can be computed. However, two observations are in order.

1. Assuming the same performance index and the same information (say full states can be observed), then each algorithm should give identical closed-loop poles and identical tracking performance. This is obvious because if one minimises the same performance index with the same d.o.f., the optimum control increment must be the same. Assuming full information and no uncertainty, each model will give identical predictions and hence the performance indices will be equivalent.

2. The closed-loop pole polynomial tends to have as many non zero poles as there are open-loop process poles (including the integrator)[¶]. If this is not true, there is a good chance you have a bug in your code. You will note that the order of P_c arising from, for instance eqn.(4.36) appears to be greater than this; in this case polynomial P_c will have several zero coefficients.

Summary: When coding, you should check for errors by ensuring consistency of behaviour in the nominal case, regardless of the model adopted.

[¶]This conclusion is obvious from (4.48) which shows that the number of poles is given by the dimension of the closed-loop state matrix.

4.8 Constraint handling

It was stated in optimisation (4.15) that MPC can deal, systematically, with constraints. However, this chapter has so far ignored that rather important point. This was largely to illustrate that in the constraint free case a fixed term control law arises, and hence one can do stability and sensitivity analysis. It is worthwhile doing a proper analysis of the unconstrained problem because if the control law does not give good performance and robustness in the unconstrained case then one cannot expect good performance in the constrained case. Unfortunately the opposite does not hold true so that good performance of the unconstrained loop does not need to imply good performance of the constrained loop; we shall consider this issue in more detail in Chapter 8.

In this section the aim is to demonstrate how optimisation (4.15) is set up and hence solved.

4.8.1 The constraint equations

The constraints may occur on any variables in the loop. The desire is that none of the constraints are violated by the optimal predictions. Hence the normal procedure is to compare the predictions with the constraints over the horizons n_y, n_u in the performance index J. The following subsections show how to set up the equations for these comparisons for common constrained variables. One can easily extend this methodology to other variables if required.

Key point: The constraint equations should be expressed in terms of the d.o.f., in this case $\underset{\rightarrow}{\Delta \mathbf{u}}$. One can then select the d.o.f. to ensure constraint satisfaction.

4.8.1.1 Input rate constraints

Take upper and lower limits on the input rate to be

$$\underline{\Delta \mathbf{u}} \leq \Delta \mathbf{u} \leq \overline{\Delta \mathbf{u}} \tag{4.80}$$

Given that the input increments are predicted to be zero beyond the horizon n_u, one can check the constraints up to then with the following:

$$\underbrace{\begin{bmatrix} \underline{\Delta \mathbf{u}} \\ \underline{\Delta \mathbf{u}} \\ \vdots \\ \underline{\Delta \mathbf{u}} \end{bmatrix}}_{\underline{\Delta U}} \leq \begin{bmatrix} \Delta \mathbf{u}_k \\ \Delta \mathbf{u}_{k-1} \\ \vdots \\ \Delta \mathbf{u}_{k+n_u-1} \end{bmatrix} \leq \underbrace{\begin{bmatrix} \overline{\Delta \mathbf{u}} \\ \overline{\Delta \mathbf{u}} \\ \vdots \\ \overline{\Delta \mathbf{u}} \end{bmatrix}}_{\overline{\Delta U}} \tag{4.81}$$

or in simpler terms

$$\underline{\Delta U} \leq \underset{\rightarrow}{\Delta \mathbf{u}} \leq \overline{\Delta U} \tag{4.82}$$

A more conventional representation for this is in terms of a single set of linear inequalities, for instance

$$\begin{bmatrix} I \\ -I \end{bmatrix} \underset{\rightarrow}{\Delta \mathbf{u}} - \begin{bmatrix} \overline{\Delta U} \\ -\underline{\Delta U} \end{bmatrix} \leq 0 \tag{4.83}$$

where I represents a suitable dimension identity matrix.

4.8.1.2 Input constraints

First one must express the future inputs in terms of the d.o.f. $\underset{\rightarrow}{\Delta \mathbf{u}}$. It is easy to see that

$$\underset{\rightarrow}{\mathbf{u}} = C_{I/\Delta} \underset{\rightarrow}{\Delta \mathbf{u}} + \underbrace{\begin{bmatrix} I \\ I \\ \vdots \\ I \end{bmatrix}}_{L} \mathbf{u}_{k-1} \tag{4.84}$$

(Recall that $C_{I/\Delta}$ is a Toeplitz matrix based on $I/(1 - z^{-1})$ and hence is lower triangular with identity matrices filling the lower triangular part.)

Take upper and lower limits on the input

$$\underline{\mathbf{u}} \leq \mathbf{u} \leq \overline{\mathbf{u}} \tag{4.85}$$

One can test for satisfaction of these constraints over the prediction horizon n_u (this automatically gives satisfaction thereafter, as the predictions assume that $\mathbf{u}_{k+n_u+i} = \mathbf{u}_{k+n_u-1}, \forall i \geq 0$) with the following:

$$\underbrace{\begin{bmatrix} \underline{\mathbf{u}} \\ \underline{\mathbf{u}} \\ \vdots \\ \underline{\mathbf{u}} \end{bmatrix}}_{\underline{U}} \leq \begin{bmatrix} \mathbf{u}_k \\ \mathbf{u}_{k+1} \\ \vdots \\ \mathbf{u}_{k+n_u-1} \end{bmatrix} \leq \underbrace{\begin{bmatrix} \overline{\mathbf{u}} \\ \overline{\mathbf{u}} \\ \vdots \\ \overline{\mathbf{u}} \end{bmatrix}}_{\overline{U}} \tag{4.86}$$

Hence, substituting from (4.84)

$$\underline{U} \leq C_{I/\Delta} \underset{\rightarrow}{\Delta \mathbf{u}} + L\mathbf{u}_{k-1} \leq \overline{U} \tag{4.87}$$

Rearranging into a more conventional form gives

$$\begin{bmatrix} C_{I/\Delta} \\ -C_{I/\Delta} \end{bmatrix} \underset{\rightarrow}{\Delta \mathbf{u}} - \begin{bmatrix} \overline{U} - L\mathbf{u}_{k-1} \\ -\underline{U} - L\mathbf{u}_{k-1} \end{bmatrix} \leq 0 \tag{4.88}$$

4.8.1.3   Output constraints

Output constraints can be set up analogously to input constraints by defining: (i) the dependence on the d.o.f. and (ii) the constraint limits – here we shall use $\underline{Y}$, $\overline{Y}$. For illustration, using the predictions (3.21), the corresponding linear inequalities are:

$$\begin{bmatrix} H \\ -H \end{bmatrix} \Delta \underset{\rightarrow}{\mathbf{u}} - \begin{bmatrix} \overline{Y} - Q\Delta \underset{\leftarrow}{\mathbf{u}} - P\underset{\leftarrow}{\mathbf{y}} \\ -\underline{Y} - Q\Delta \underset{\leftarrow}{\mathbf{u}} - P\underset{\leftarrow}{\mathbf{y}} \end{bmatrix} \leq 0 \qquad (4.89)$$

Note that the ouput constraints are taken up to the prediction horizon n_y. In fact as will be discussed in the chapter on feasibility, one should ordinarily take the horizon even bigger than this, as non-inclusion in the cost function J does not imply that the outputs are no longer changing. It could be that the unconsidered part of the output prediction, beyond the horizon n_y, does violate a constraint. Failure to account for this may not be resolved simply by the usual receding horizon arguments.

4.8.1.4   Summary

All the constraints must be satisfied simultaneously; hence one can combine equations (4.83, 4.88, 4.89) into a single set of linear inequalities of the form:

$$C\Delta \underset{\rightarrow}{\mathbf{u}} - \mathbf{d}_k \leq 0 \qquad (4.90)$$

where

$$C = \begin{bmatrix} I \\ -I \\ C_{I/\Delta} \\ -C_{I/\Delta} \\ H \\ -H \end{bmatrix} \; ; \quad \mathbf{d} = \begin{bmatrix} \overline{\Delta U} \\ -\underline{\Delta U} \\ \overline{U} - Lu_{k-1} \\ -\underline{U} - Lu_{k-1} \\ \overline{Y} - Q\Delta \underset{\leftarrow}{\mathbf{u}} - P\underset{\leftarrow}{\mathbf{y}} \\ -\underline{Y} - Q\Delta \underset{\leftarrow}{\mathbf{u}} - P\underset{\leftarrow}{\mathbf{y}} \end{bmatrix}$$

$\mathbf{d}_k$ depends upon past input and output information and C is time invariant.

The reader will note that even with a relatively small n_u and n_y, nevertheless there can be a very large number of linear inequalities. Let the system be $n \times n$; then the constraints above give p inequalities where

$$p = 2n(2n_u + n_y) \qquad (4.91)$$

Summary: Constraints at each sampling instant can be represented by a set of linear inequalities (4.90) that depend upon the d.o.f. and the current state. Even low dimensional systems can require large numbers of inequalities to be handled.

4.8.2 Solving the constrained optimisation

We shall leave a more detailed discussion of this until later. Substituting from (4.90) and, for instance (4.21), optimisation (4.15) takes the form

$$\min_{\underset{\rightarrow}{\Delta \mathbf{u}}} \; J = \underset{\rightarrow}{\Delta \mathbf{u}}^T S \underset{\rightarrow}{\Delta \mathbf{u}} + 2 \underset{\rightarrow}{\Delta \mathbf{u}}^T \mathbf{f} \;\; \text{s.t.} \;\; C \underset{\rightarrow}{\Delta \mathbf{u}} - \mathbf{d}_k \leq 0 \qquad (4.92)$$

where $S = H^T H + \lambda I$, $\mathbf{f} = [P \underset{\leftarrow}{\Delta \mathbf{u}} + Q \underset{\leftarrow}{\mathbf{y}} - \underset{\rightarrow}{\mathbf{r}}]$. This is known as a quadratic programming (QP) problem for which solvers are easy to find. Most industrial suppliers of MPC write their own solvers and indeed the topic is still open to new ideas in the academic literature. Some discussion of this can be found in [13, 78]. Here it is suffice to say:

1. Solving a high dimensional QP on-line is considered tractable in the process industry where time constants are not fast.

2. MATLAB supplies a QP solver which will often compute the answer with 10 d.o.f. or more in well under a second.

3. The conventional approach is the active set method. This has guarantees of convergence and is fairly simple to code; the theoretical limits on the number of iterations is very large but rarely approached in practice.

4. Interior point methods are becoming more popular, as are multiparametric methods.

Summary: The on-line optimisation reduces to a QP for which standard solvers exist. It is considered a tractable problem.

4.8.3 Hard and soft constraints

In practical problems it is common to find that the desirable constraints are inconsistent; that is, they cannot all be satisfied simultaneously. In the case that constraints (4.90) do not admit a solution, then the MPC optimisation is ill posed and has no solution. Clearly one cannot afford for this to happen on a real process as the resulting control would be arbitrary. Each process will build in its own fail safes for such an occurrence but a more generic procedure does exist.

Constraints should be placed into two categories:

1. Hard limits – e.g. valves cannot go beyond fully open or fully shut.

2. Soft limits – these can be violated though at some penalty such as loss of product quality and safety values blowing.

The soft limits should then be further prioritised into a ranking of importance. In the event that constraints (4.90) cannot be satisfied, then one could gradually relax the soft constraints in the given priority order until a feasible solution exists.

The topic of how and which constraints to relax is a little open ended and will be very process dependent. Based on the process given, appropriate algorithms largely exist in the literature but these are more in the realm of the numerical analyist than the control engineer so are not given here. The simplest approach is to choose the control that gives the least maximum weighted constraint violation (e.g. [101]) over all the soft constraints; that is, minimise an infinity norm, e.g.

$$\min_{\Delta \underrightarrow{\mathbf{u}}} \quad W \| C \Delta \underrightarrow{\mathbf{u}} - \mathbf{d}_k \|_\infty \tag{4.93}$$

where W is a weighting matrix ($W_{i,i}$ is infinite for hard constraints and 0 for rows with negative values – satisfied constraints). Standard linear programming algorithms using slack variables or Lawson's algorithm [101] (a form of interior point method) will solve this problem.

> **Summary:** In the event of infeasibility of constraints, the soft constraints must be relaxed to ensure the MPC algorithm has a well-posed optimisation. This procedure is nongeneric and may often be dealt with at a higher (supervisory) level rather than in the MPC algorithm. A more detailed discussion is given in Chapter 8.

4.8.4 Stability with constraints

Although one can define an MPC algorithm that does constraint handling, it is not straightforward to establish if that algorithm will be stabilising. The stability of the underlying linear controller (e.g. 4.31, 4.47) does not guarantee the stability of the constrained controller, at least in the finite horizon case which is typical in practice.

There is no easy answer to this failing; however, the lack of a guarantee of stability does not imply instability. In practice, if sensible guidelines (use a large enough output and constraint horizon) are followed and one avoids infeasibility, then the constrained loop will have a similar performance to the unconstrained loop. This topic is considered in more detail in the Chapters 6 and 8.

> **Summary:** Practical industrial algorithms with constraint handling do not have stability guarantees, but this is rarely a problem.

4.9   Simple variations on the basic algorithm

4.9.1   Alternatives to the 2-norm

In the late 1980s and early 1990s there was some investigation into the use of 1-norms and ∞-norms (e.g. [1], [106], [159]) in the performance index as opposed to the 2-norm. This amounts to minimising the worst case error or the sum of the error moduli. Such a change facilitated better a priori robustness results for certain classes of model uncertainty and moreover a reduction in computational load was possible as the optimisation implied was a linear program (LP) rather than quadratic program (QP). However, the typical control was not smooth and therefore such approaches were never really accepted as attractive in practice. This area is open to further study.

4.9.2   Alternative parameterisations of the degrees of freedom

Classical MPC algorithms use the future control increments as the d.o.f. in the optimisation. Although in practice this works quite well for n_u large enough, it could be argued that it is inefficient and not always numerically well conditioned. It is possible to use other parameterisations. There are many possibilities such as prestabilisation [60]; closed-loop prediction [59],[66], [129]; Laguerre functions [154]; non-symmetric spacing of control changes and PFC [98]. In truth this area is under-researched and the author is unaware of any good summaries. Significant improvements in performance or reductions in computation are possible with a change in the parameterisation of the d.o.f. so some possibilities are discussed later and the concept is implicit in Chapter 6.

4.9.3   Improving response to measurable disturbances

In many process there are measurable disturbances with a known effect on the output predictions. In this case one could improve closed-loop response significantly by incorporating this knowledge into the prediction equations. For instance, if the process took the form

$$D(z)\mathbf{y} = N(z)\mathbf{u} + \frac{\zeta}{\Delta} + c(z)f \tag{4.94}$$

where $c(z)$ is known and f is a measurable disturbance, then it is straightforward to rederive the prediction equations to include f. The resulting predictions would take a form similar to (3.21), e.g.

$$\underset{\rightarrow k}{\mathbf{y}} = H\Delta\underset{\rightarrow k-1}{\mathbf{u}} + P\Delta\underset{\leftarrow k-1}{\mathbf{u}} + Q\underset{\leftarrow k}{\mathbf{y}} + F\begin{bmatrix} \underset{\leftarrow}{\Delta f} \\ \underset{\rightarrow}{\Delta f} \end{bmatrix}; \quad F = C_D^{-1}\begin{bmatrix} H_c \\ C_c \end{bmatrix} \tag{4.95}$$

Clearly substituting (4.95) into (4.6) gives the optimum control law as

$$\Delta\underset{\rightarrow}{\mathbf{u}} = [H^T H + \lambda I] H^T [\underset{\rightarrow}{\mathbf{r}} - P\Delta\underset{\leftarrow}{\mathbf{u}} - Q\underset{\leftarrow}{\mathbf{y}} - F \begin{bmatrix} \Delta\underset{\leftarrow}{f} \\ \Delta\underset{\rightarrow}{f} \end{bmatrix}] \tag{4.96}$$

Although one could argue that, at least in steady state, the integrator will deal with all disturbances, it is important to remember that the more accurate the predictions, the better the control is likely to be!

> **Summary:** If you have any information that can be used to improve prediction accuracy, use it! Performance is directly linked to prediction accuracy.

4.10 Predictive functional control (PFC)

This algorithm is discussed more fully in Chapter 13 and so only a summary is given here.

In PFC use is made of coincidence points. That is, a few points in the future are selected and the predicted errors are considered only at these rather than over an entire horizon. The optimisation is replaced by equality conditions, hence simplifying computation considerably. In practice it can give a very similar performance to MPC with a much simpler and faster algorithm which thus can be implemented cheaply and on fast processes. However, not much has appeared in the literature and this algorithm is justified mainly by successful industrial implementation.

> **Summary:** Until now applications of PFC have largely been restricted to the SISO case. On many applications the performance is very similar to that achievable with a full GPC algorithm and this at fraction of the complication and computational load.

4.10.1 Predictive functional control with one coincidence point

A typical algorithm with just one coincidence point would set a target trajectory (n-steps ahead) such as

$$\text{target} = \mathbf{y}_k + (\mathbf{r} - \mathbf{y}_k)(1 - \beta^n) \tag{4.97}$$

That is, the target assumes the response of a first order lag with pole β moving from the current to the desired output. The designer selects the pole β to be close to the desired closed-loop response.

However, instead of using the whole trajectory, PFC selects just one point n_y steps ahead, where n_y is denoted as the coincidence horizon. Let the corresponding prediction be

$$\mathbf{y}_{k+n_y} = \mathbf{e}_{n_y}^T \underset{\rightarrow}{\mathbf{y}} \tag{4.98}$$

PFC then uses the d.o.f. in $\underset{\rightarrow}{\mathbf{y}}$ (usually just one control move is assumed free, hence $\Delta\underset{\rightarrow}{\mathbf{u}} = \Delta\mathbf{u}_k$) to satisfy the equality

$$\begin{aligned} \mathbf{y}_{k+n_y} &= \text{target} \\ \mathbf{e}_{n_y}^T \underset{\rightarrow}{\mathbf{y}} &= \mathbf{y}_k + (\mathbf{r} - \mathbf{y}_k)(1 - \beta^{n_y}) \end{aligned} \tag{4.99}$$

where $\mathbf{e}_{n_y}$ is the n_y^{th} standard basis vector. Equality (4.99) has a unique solution which can be written down by inspection. For instance, taking predictions (3.21), the computation (4.99) reduces to

$$\mathbf{e}_{n_y}^T [H\Delta\mathbf{u}_k + P\Delta\underset{\leftarrow}{\mathbf{u}} + Q\underset{\leftarrow}{\mathbf{y}}] = \beta^{n_y} \underset{\leftarrow}{\mathbf{y}} + (1 - \beta^{n_y})\mathbf{r} \tag{4.100}$$

This can easily be rearranged into the same form as (4.31) with

$$D_k = [1 + \frac{z^{-1}\mathbf{e}_{n_y}^T P}{\mathbf{e}_{n_y}^T G}]; \quad N_k = \frac{\mathbf{e}_{n_y}^T Q - \beta^{n_y}}{\mathbf{e}_{n_y}^T G}; \quad P_r = \frac{1 - \beta^{n_y}}{\mathbf{e}_{n_y}^T G} \tag{4.101}$$

Summary: The structure of a PFC control law is the same as other common predictive control laws. This depends solely on the model used to form the predictions.

4.10.2 PFC tuning parameters

Tuning of the coincidence horizons and β is by trial and error. However, one can usually estimate β very quickly as it corresponds to the desired (realistic) closed-loop dynamic. One then does a search over n_y to find the coincidence horizon giving the best performance. Such a search involves computation of controllers as in (4.101) and hence is very fast.

Summary: PFC often gives a similar performance to GPC with far simpler algorithm and trivial on-line computation. For SISO systems it is worth considering it first.

4.10.3 PFC with two coincidence points

Where the set point has ramp characteristics, the d.o.f. are parameterised in terms of a step in control at sample k and a ramp rate in the control thereafter. Equality conditions then result from the selection of two coincidence horizons. This facilitates the tracking of ramps with no lag but the tuning (essentially a global search) is less straightforward. Nevertheless, its simplicity and success in practice are still major justifications for this algorithm.

4.10.4 Limitations and summary

PFC is not really intended for multivariable plants which is one of the major strengths of other MPC algorithms, hence it does not merit a large part in this book. Moreover its major strength, that is its simplicity and hence appeal to engineers who find MPC hard to follow, is also a weakness for processes with more complex dynamics. For some unstable processes and indeed some stable processes it is not straightforward and perhaps not even possible [133] to get good performance from PFC. MPC algorithms are better able to handle complexity because they make more use of model information, the price of course being a higher load and a less transparent algorithm.

4.11   Other performance indices

Practical algorithms often use a slightly more involved cost of the form

$$J = \sum_{i=1}^{n_y} \|r_{t+i} - y_{t+i}\|_2^2 + \sum_{i=0}^{n_u-1} \lambda \|\Delta u_{t+i}\|_2^2 + \sum_{i=0}^{n_u-1} \lambda \|u_{t+i} - u_{ss}\|_2^2 \qquad (4.102)$$

where u_{ss} is the predicted steady-state input.

Following through the algebra the reader will see that this performance index still reduces to a simple quadratic in the future control moves and hence the implementation complexity is equivalent. There are, however, more weights to tune.

In fact readers will have understood by now that within certain limits they can choose whatever performance index they please. One of the major advanatages of MPC is its flexibility; users can set it up and tune it for their own specific needs and often with very little effort beyond the basic algorithm.

Summary: MPC is flexible in terms of the models and performance indices that it can utilise.

5

Examples – tuning predictive control and numerical conditioning

This chapter gives a a series of examples and demonstrates how MPC might be tuned quickly and effectively. The aim is to give the reader some insight into what effects the tuning parameters have. It is included solely for completeness and has been kept brief, as the information is already available in other books on the topic. In particular however, you should use the insight gained in this chapter to give a better understanding as to why there has been a move in the academic literature towards infinite horizons. As you read this chapter, concentrate on insight.

The chapter is divided into stable and unstable processes because as is well known very different rules are needed for these. For instance, stable processes can always be stabilised, albeit with small bandwidth, by low gain control. Unstable processes on the other hand need a sufficiently high gain to be stabilised and moreover are usually conditionally stable.

A summary of the observations is given in Section 5.7 for the reader who wants simply to scan the chapter.

5.1 Matching closed-loop and open-loop behaviour

This section gives a quick overview of what will be observed in this chapter.

If one is to expect good closed-loop behaviour then it is necessary that minimising J gives a control trajectory, in particular the first increment, that is close to the closed-loop optimal. This means that the prediction class used must include at least one member that is close to the optimal closed-loop trajectories. If the prediction class used to minimise J does not include a solution near to the desired closed-loop behaviour, then the optimisation is ill posed; that is, one is minimising something which has little bearing on what one really wants and as such control can become arbitrary and at best give poor performance.

We will return to this theme throughout the book and in particular in Chapter 6, as it is central to producing a well-posed MPC algorithm. For this chapter the focus

TABLE 5.1

Notation used in Figures 5.1–5.3

$n_y = 3$	$n_y = 5$	$n_y = 10$	$n_y = 20$
Solid	Dotted	Dashed	Dash-dot

will be simply on illustrating the performance that arises from various choices of the tuning parameters. The link to how well posed the optimisation is will be made retrospectively.

Summary: One would not expect MPC to give good control if the basic setup of the optimisation objective and d.o.f. was poor.

5.2 Single-input/single-output examples

For simplicity a single example is taken to illustrate the effects of changing the tuning parameters. The reader should note that while the effects may vary from one example to another, the main trends would be expected to be similar. You will gain a better appreciation only by doing your own designs on several examples. The reader is also reminded that the choice of underlying prediction model (and therefore control law) should not affect *nominal* performance. Hence only one of the control laws of the previous chapter will be used for illustration.

The following section implements the control law of (4.25) on a system given as

$$y = \frac{z^{-1} + .2z^{-2}}{(1 - 0.9z^{-1})(1 - 0.8z^{-1})} u \tag{5.1}$$

The simulations given are the closed-loop response to a unit step in the set point for various selections of tuning parameters. A summary of the results is given after the simulations have been displayed. No advance information of set point changes is assumed so $P_r(z)$ is replaced by the appropriate scalar.

5.2.1 Effect of varying the output horizon

Figures 5.1–5.3 show the responses for $n_u = 1$, 2, 5 respectively, when the output horizon is varied. The notation used for plots represented different n_y is detailed in Table 5.1.

5.2.2 Effect of varying the input horizon

Figure 5.4 shows the responses for $n_y = 10$ for different n_u as detailed in Table 5.2.

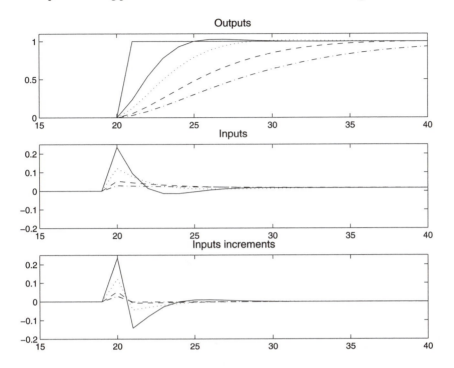

FIGURE 5.1

Responses for varying n_y with $n_u = 1$.

TABLE 5.2

Notation used in Figure 5.4

$n_u = 1$	$n_u = 2$	$n_u = 3$	$n_u = 5$
Solid	Dotted	Dashed	Dash-dot

5.2.3   Effect of varying the control weighting

Figure 5.5 shows the responses for $n_y = 10$, $n_u = 3$ for different λ as detailed in Table 5.3.

5.2.4   Summary

It is noted that the impact of changes to an individual horizon depends upon how the other horizon is selected. So for instance:

- If n_u is small (Figure 5.1) increasing n_y causes the loop dynamics to slow down.

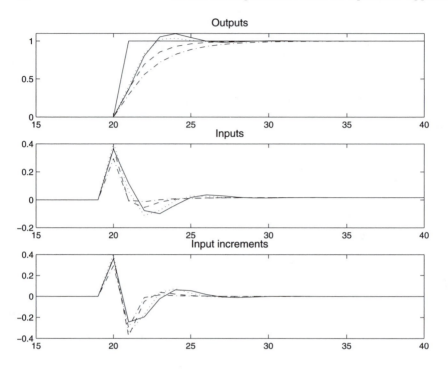

FIGURE 5.2

Responses for varying n_y with $n_u = 2$.

TABLE 5.3

Notation used in Figure 5.5

$\lambda = .1$	$\lambda = 1$	$\lambda = 10$	$\lambda = 100$
Solid	Dotted	Dashed	Dash-dot

- If n_u is large (Figure 5.3), increasing n_y improves performance.

- If n_y is large (Figure 5.4), then increasing n_u improves performance.

- If n_y is small, then increasing n_u can lead to near deadbeat behaviour.

- Increasing λ slows down the responses.

5.2.4.1   Best choice of horizons

From the above observations one could make the following inferences:

1. As n_y is increased, nominal closed-loop performance improves if n_u is large enough.

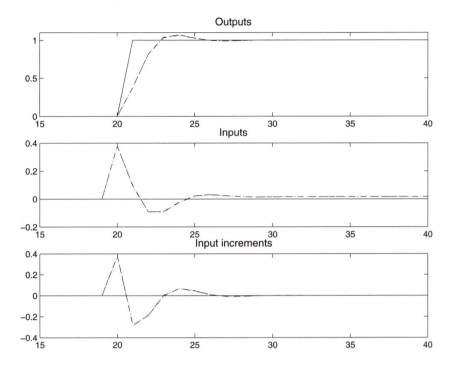

FIGURE 5.3

Responses for varying n_y with $n_u = 5$.

2. As n_u is increased nominal closed-loop performance improves if n_y is large enough. However, for many models, there is not much change beyond $n_u = 3$.

Summary: It seems that for many processes, one can optimise expected closed-loop performance by choosing n_y and n_u to be as large as possible.

5.2.4.2 Choice of horizons n_y, n_u to ensure a well-posed optimisation

These example responses have also demonstrated the need to give a good match between the open-loop prediction class and the closed-loop behaviour that is desired. For instance:

- With n_y large and $n_u = 1$ (dash-dot line of Figure 5.1), the minimisation has only one control move available and hence chooses the move that is likely to eliminate steady-state offset; this is often called mean-level control. Hence at best one will get open-loop dynamics.

- A better response (Figure 5.3) arises where more control changes are allowed in transients and hence the open-loop predictions can be closed to the desired

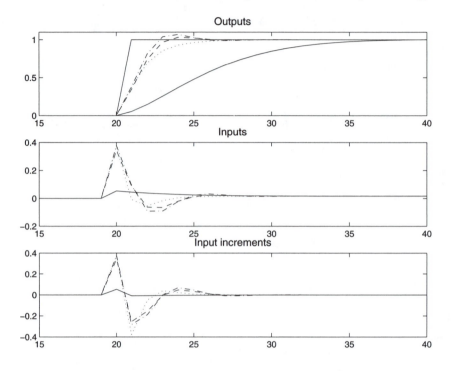

FIGURE 5.4

Responses for varying n_u with $n_y = 10$.

closed-loop behaviour. Unsurprisingly larger n_u results in better closed-loop performance.

However, it should be emphasised that the conclusions of Section 5.2.4.1 are incomplete. They only apply where $n_y \gg n_u$ and this is because of arguments about mismatch between the predictions and the desired behaviour. If $n_y \approx n_u$, then a substantial amount of transient behaviour (that beyond n_y) is not included within the cost function and this could imply a poorly defined optimisation as one is only optimising part of the transient behaviour, regardless of the impact on the ignored part and hence on the actual closed-loop behaviour. Hence, with $n_y \approx n_u$, it is difficult to say a priori whether the resulting controller will give good behaviour. This issue is discussed more fully in Chapter 6.

Summary: The optimisation maybe ill posed unless the following are true:

1. n_u should be large.

2. $n_y - n_u$ should be greater than the settling time.

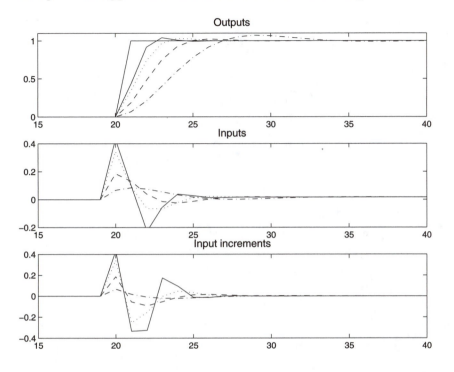

FIGURE 5.5
Responses for varying λ with $n_u = 3$, $n_y = 10$.

5.2.4.3   Choice of λ

The effect of changes to control weighting (Figure 5.5) seems fairly clear; that is, the more the weighting the less active the input changes are. However, one should take this conclusion with some caution, as it is also subject to the remarks of Section 5.1; that is, how well posed is the performance index?

Try the example of (5.1) with $n_u = 1$, $n_y = 40$ and $\lambda = 1$, 100, 1000, 10000 and plot the corresponding step responses in Figure 5.6. Surprisingly perhaps, the responses vary very little despite the large changes in λ. In fact one can only just begin to notice a difference once $\lambda = 10000$ which is several orders of magnitude greater than that required to make the changes observed in Figure 5.5. One might wonder: how effective is λ as a tuning parameter?

There is a need for some simple insight. The reason for the apparent ineffectiveness of λ is that the performance index is dominated by the steady-state output tracking error. So the optimisation places all the emphasis on making tracking errors small and the impact of the control weighting is relatively small. However, if one changed the output horizon to, for instance, $n_y = 8$, the control weighting now has a large

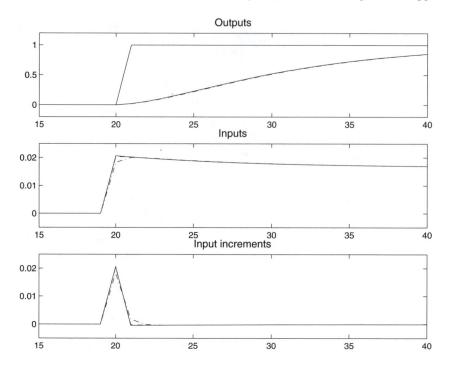

FIGURE 5.6

Responses for varying λ with $n_y = 40$, $n_u = 1$.

impact, as the cost is no longer dominated by tracking errors.

In conclusion, control weighting has an impact where the magnitude of the control weighting terms $\lambda \|\Delta \underset{\rightarrow}{u}\|_2^2$ is of similar or larger magnitude than the tracking error term. Hence the required λ is closely linked to the input and output horizons. The typical guideline of normalising signals and then making all the weights unity is a good default only where n_u is large.

Summary: λ is an effective tuning parameter only for the ranges of λ where the gradient (w.r.t. the d.o.f.) of the term $\lambda \|\Delta \underset{\rightarrow}{u}\|_2^2$ is similar enough in magnitude to the gradient of the tracking error term.

5.3   The benefits of systematic constraint handling

One of the advantages of MPC is the ability to incorporate constraints into the optimisation. This advantage is demonstrated in this section by way of a simple simulation example and some simple discussion.

More traditional strategies such as PID may use integral desaturation and other techniques to deal with constraints, but in essense these reduce to the statement: *if the controller demands a control action that exceeds constraints, then replace it with a control action at the constraint.* MPC however, because it can include knowledge of the constraints within the optimisation, can be far more intelligent and hence give large improvements in performance where constraints are active.

5.3.1 Simple example of weakness in a saturation policy

Consider the case where you are driving a car around a corner. The PID controller takes the curvature of the road and the limits of the car to set a desired speed, say 40 mph. However, the PID controller does not look farther down the road and notice constraints on *future* behaviour such as: (i) an impending tighter curvature, (ii) a parked car to be avoided or (iii) a rather large pothole. Hence, when these factors become 'present', it is too late to take evasive action, as the car is already operating at its limit and a crash follows.

MPC on the other hand looks at the behaviour over a sensible future horizon, say 100 yds ahead. As such its strategy takes account of *future* constraints and thus takes the car along the road more cautiously so that there is sufficient control freedom remaining to deal with the constraints when they arise.

The key point to notice is that current behaviour is affected by future constraints. If one ignores future constraints, then the current behaviour maybe unwise.

5.3.2 Numerical illustration of the weaknesses in saturation policies

Next an illustration is given of how much improvement in performance systematic constraint handling can give as opposed to a simple saturation policy, that is, one where if the input exceeds a limit, then one simply implements the input at that limit. Consider the process

$$y = \frac{0.1z^{-1} - 0.2z^{-2}}{1 - 1.5z^{-1}} u \tag{5.2}$$

and include input limits

$$-1.7 \leq \Delta u \leq 2; \quad |\Delta u| \leq 1.4 \tag{5.3}$$

Take the tuning parameters to be $n_y = 10, n_u = 3, \lambda = 0.02$ and find the closed-loop step responses. These are displayed in Figure 5.7 where the dotted line is for the constrained MPC and the solid line represents a strategy using simple saturation. One can see clearly that, even though there seems to be very little difference in the input plots, this difference is enough to make the closed-loop output simulation unstable

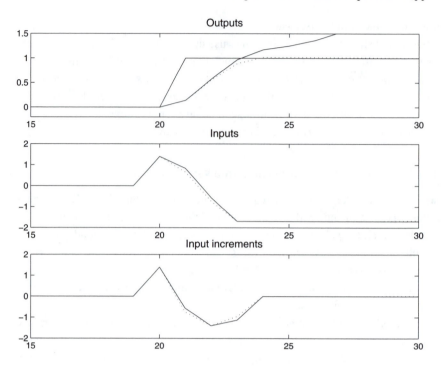

FIGURE 5.7

Responses in the presence of constraints.

for the saturation algorithm (solid line). That is, failure to take proper account of the constraints during the control calculations has led to disaster, i.e. instability.

The proper constrained algorithm proposed trajectories satisfying constraints over the whole future horizon and hence did not overstretch itself; hence the corresponding responses are good. The saturation algorithm simply chooses the current action without consideration of the impact over the future with possibly dire consequences for performance.

Several papers and books have appeared (e.g. [13], [149]) discussing this issue and looking at the implied quadratic programming problem in more detail. Hence that discussion is not repeated here. It is clear that at times saturation gives the same result as a proper constrained optimisation; however, it is also clear that many times it does not, especially for the MIMO case and where there are state constraints. Failure to include constraints can have serious consequences for performance and even stability. In fact the arguments go back to those of Section 5.1 – in order to give a well-posed optimisation one must ensure a good match between the predictions and the resulting closed-loop behaviour. At the very least this implies that the prediction class allowed within the optimisation must satisfy constraints over the entire future; these are often denoted as *feasible* predictions*.

Summary: If constraints are present, these should usually be included explicitly into the optimisation of predicted performance or in other words one should optimise over a *feasible* prediction class.

Simple saturation strategies may work well on tame examples but should be used with extreme caution.

5.4 Unstable examples

In theory predictive control should deal well with unstable systems. However, in this section one will observe a conclusion that seems to contradict that given for stable systems. That is, if one increases the output horizon beyond a certain point, performance deteriorates rather than improves. Moreover, the use of $n_u = 1$ is often incompatible with good performance, unlike for stable systems.

Readers will be aware that unstable systems are usually conditionally stable; that is, the gain must lie between an upper and lower limit. In a similar vein, conventionally designed MPC controllers of unstable systems are conditionally stabilising in that the output horizon must lie between an upper and a lower limit. This will be demonstrated next.

5.4.1 Effects of tuning parameters on unstable systems – a counter intuitive result

Take the unstable example of (5.2). Plot the closed-loop responses for $n_u = 1$, $n_y = 2, 6, 10$ and 40 (solid, dotted, dashed and dash-dot lines, respectively) in Figure 5.8 and $n_u = 2$, $n_y = 2, 6, 10$ and 40 in Figure 5.9. In each case let $\lambda = 1$.

It is clear that increasing n_u from 1 to 2 has improved performance dramatically; the reason for this will be explained in more detail in a later chapter. However the following are more interesting:

- For $n_u = 1$ the system is unstable for low n_y and the closed-loop is arbitrarily slow for high n_y.

- For $n_u = 2$ the system is unstable for low n_y (solid line), performs well for intermediate n_y and then is beginning to lose performance for high n_y (dash-dot line).

---

*These are discussed in more detail in Chapter 8.

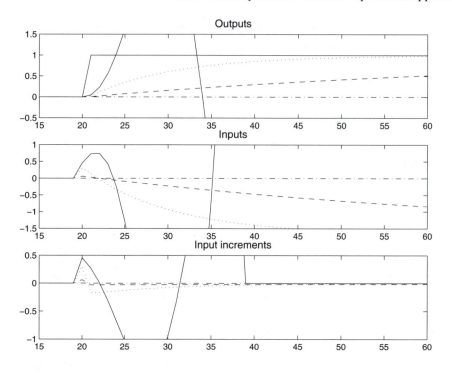
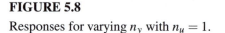

FIGURE 5.8
Responses for varying n_y with $n_u = 1$.

- In both cases performance deteriorates with high n_y.

As it happens with $n_u = 1$ the optimisation is simply ill posed for the reasoning given in Section 5.1. That is, with only one d.o.f. the open-loop predictions cannot both be stabilised and achieve the target set point; one d.o.f. is needed to stabilise the unstable dynamics and another to achieve a given target. Hence as n_y increases, the focus of the optimisation is ensuring that the unstable open-loop predictions do not diverge to a large number which implies that the current input increment must be very small, hence resulting in very slow closed-loop dynamics.

With $n_u = 2$ there is sufficient d.o.f. to achieve both requirements, hence there is a dramatic improvement in performance. However, now as n_y becomes larger the the solution loses accuracy, hence the deterioration. The reader might be wondering how this can be, as logically increasing n_y means that one has captured more of the predictions, hence there should be a better match between predictions and actual behaviour, implying an improvement in performance. The explanation and corresponding warning is given in the next section and a suitable solution is given in a later chapter.

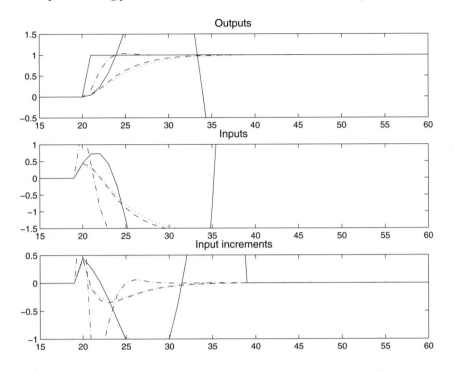

FIGURE 5.9
Responses for varying n_y with $n_u = 2$.

Summary: Use of high n_y in a *standard* MPC design may cause severe performance degradation for an open-loop unstable plant. This seems counter intuitive given the insights of Section 5.2.4.1.

For unstable processes it may be difficult to obtain good performance with $n_u = 1$. A typical guideline is that $n_u \geq p + 1$, where p is the number of unstable poles.

5.5 Numerical ill-conditioning with open-loop unstable systems

The basic difficulty with MPC of unstable systems is that the optimisation of (4.4) is ill-conditioned for large n_y and ill posed for small n_y. Hence if n_y is small the minimisation may not give a sensible answer due to ignored transient behaviour and instability can result. However, if n_y is large, the ill-conditioning will cause an erroneous solution to the optimisation and again instability may result.

We shall not dwell on the use of small n_y, as this is obviously ill posed. If one minimises over a small prediction horizon, one is not taking into account behaviour beyond that. If that behaviour is divergent, as the prediction equations would suggest, then to ignore it is foolish. MPC works on the basis that the part of the predictions that are ignored (beyond the output horizon) should be largely in steady state [†]; this is clearly not the case for unstable systems.

If one takes n_y to be large there is a hope that the predictions will have settled and the unstable dynamics will not cause the predictions to diverge beyond the output horizon. Hence use of large n_y should give good performance, in principle at least. Why does this expectation fail then for large n_y?

5.5.1 How does ill-conditioning arise?

Consider the optimisation of (4.4). This is based on matrices H, P, Q. For an unstable system the coefficients of these matrices diverge with the row index. For instance, the elements of H include the coefficients of the system step response, which diverges. As a consequence, even though H itself is Toeplitz, it contains elements with a large variation in magnitude. Hence H, P, Q are ill-conditioned matrices.

One can demonstrate the ill-conditioning quite effectively using a graph of the controller coefficients N_k, D_k against prediction horizon n_y. Take the unstable system

$$G(z) = \frac{0.5z^{-1} + z^{-2}}{1 - 4z^{-1} + 3.75z^{-2}} \tag{5.4}$$

Sketch in Figure 5.10 the coefficients $N_k(1)$, $D_k(2)$ vs n_y for $n_u = 1, 2$ (solid and dotted lines respectively). It is clear that the coefficients converge well at first (as would be expected) but then beyond $n_y = 20$ the ill-conditioning is so bad that even with 16 decimal places of MATLAB the controller coefficients become random (there were smaller inaccuracies at lower n_y).

The value of n_y at which the ill-conditioning becomes significant is process and precision dependent and so an analysis of this is not interesting. As the optimisation is ill conditioned it is more fruitful to find means of ensuring the optimisation has better conditioning. This is done in Chapter 7.

Summary: To apply a conventional MPC algorithm to unstable systems one must choose n_y carefully, not too small and not too large. One must also use the result with caution. In fact you are better advised not to use a conventional algorithm at all and use more recent variants such as those based on the closed-loop paradigm (see Chapter 7) which have better conditioning.

Remark 5.1 *The results of this section have been illustrated with transfer function models [104]. However, similar conclusions will apply to GPC based on state-space models (see [113]). FIR models do not apply to unstable processes.*

[†]This is so that the ignored errors are essentially zero.

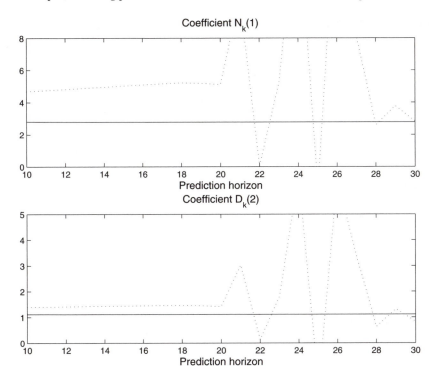

FIGURE 5.10

Coefficient values for varying n_y.

5.6 MIMO examples

This chapter contains only a brief summary of applications to MIMO examples, as most the conclusions are the same as for the SISO case. The focus is on what changes and what else the designer must bear in mind.

Due to the use of a performance index, GPC gives an optimal management of inter-action, at least by way of performance specified in J. Hence if the designer wishes to trade off performance in one loop with another, the systematic route is via the weightings in J. That is, if one increases the relative weight on one loop w.r.t. to another, then the relative errors will reduce. However, there is no simple analytic link to the actual weights, at least not when J is quadratic. So if the variance in one loop was too high, one could increase the weight on that loop but this would be a trial and error process and the repercussions on errors in the other loops would not be known a priori.

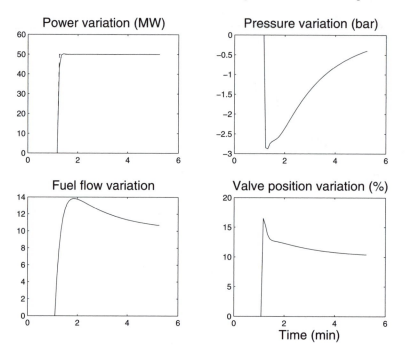

FIGURE 5.11

Responses for a MIMO system with output weights diag([1 1]).

An example is given next based on a simplified boiler model for electricity generation. This is a stiff system whose state-space matrices are given as

$$A = \begin{bmatrix} -0.0042 & 0 & 0.0050 \\ 0.0192 & -0.1000 & 0 \\ 0 & 0 & -0.1000 \end{bmatrix} ; \quad B = \begin{bmatrix} 0 & -0.0042 \\ 0 & 0.0191 \\ 0.1000 & 0 \end{bmatrix}$$

$$C = \begin{bmatrix} 1.6912 & 13.2160 & 0 \\ 0.8427 & 0 & 0 \end{bmatrix} ; \quad D = \begin{bmatrix} 0 & 1.6854 \\ 0 & -0.1568 \end{bmatrix}$$

(5.5)

Closed-loop responses for a step demand in power output are given in Figures 5.11,
5.12 for tuning parameters $n_y = 20$, $n_u = 3$ and output weights $W_y = \text{diag}([1,1])$, $W_y = \text{diag}([1,10])$, respectively. It is clear that changing the emphasis on the second output (the pressure) has changed the performance, that is the speed in this loop, at the
expense of more input activity. However, it is difficult from this to say what the
weighting should be. (We should note that a 3 bar variation is larger than would usually be tolerated.) One could reduce the peak in firing rate (fuel flow) by increasing
the weighting on this input, say to 20, which gives the response of Figure 5.13. It is
evident however, that this process is somewhat ad hoc.

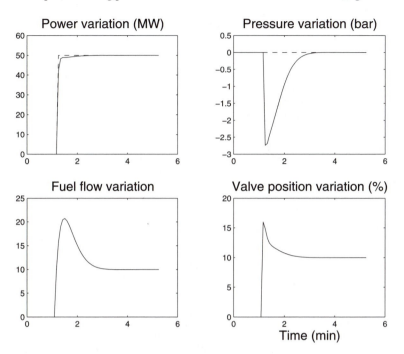

FIGURE 5.12

Responses for a MIMO system with output weights diag([1 10]).

Summary: MIMO design is linked directly to the cost function J. The predicted performance is always optimised w.r.t. J and not less mathematical concepts such as overshoot or relative bandwidth. You change the design by changing J, usually the weights.

The means of selecting weights to achieve the best by way of other than the 2-norm objective of (4.16) is not clear and can be a very time consuming part of an MPC design.

5.7 Summary of guidelines

In order to ensure that the optimisation is well posed, that is the minimisation bearing a direct relationship to the expected closed-loop behaviour, one should choose:

1. The output horizon n_y should be larger than n_u plus the system settling time. For a rigorous proof one should choose $n_y = \infty$.

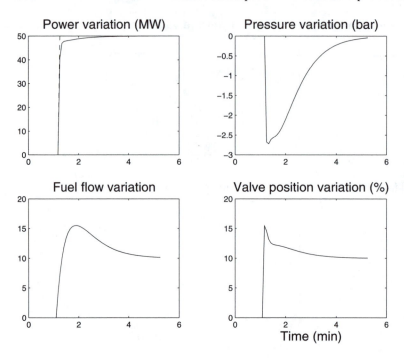

FIGURE 5.13

Responses for a MIMO system with input weights diag([1, 20]).

2. The input horizon should be as large as the expected transient behaviour. In practice a value of $n_u \geq 3$ often seems to give performance close to the '*global optimal*'. To achieve closed-loop behaviour close to open-loop behaviour, $n_u = 1$ will often be sufficient.

3. The weighting matrices can be used to shift the emphasis onto different loops but the efficacy of any change depends on the corresponding gradients in the cost function; sometimes an apparently large change in weight may have negligible effect on the optimum but this could probably be easily understood with a simple analysis of the actual objective (4.16).

4. With unstable processes one should plot[‡] a graph analogous to Figure (5.10) to ensure the answer is reliable. Alternatively use a better conditioned approach to be presented in Chapter 7.

[‡]As an alternative, do a more rigorous conditioning analysis.

6

Stability guarantees and optimising performance

It will be shown that two main components ensure[*] the existence of an a priori stability guarantee for an MPC algorithm. These are:

1. The current prediction class now must contain as a member the optimal trajectories from the previous sampling instant (denoted *the tail*).

2. The cost should contain infinite output horizons.

Furthermore to ensure good performance there are two further requirements:

3. The minimisation of prediction mismatch is needed.

4. To facilitate the proofs in the constrained case the concept of invariance (Chapter 11) is required.

The first three of these four components and the relevant discussions are introduced in this chapter.

Earlier chapters gave stability results for the constraint free case or rather showed how the implied closed-loop poles could be computed; see for instance eqns.(4.36, 4.48). Clearly however, these results are not valid when constraints are active as they are based on linear analysis. This chapter will show how a more generic approach to stability can be developed and moreover one that also applies to the constrained case (where a nonlinear analysis must be used). Nevertheless it is noted here for completeness that the proof only applies to the constrained case when there is feasibility; this concept and suitable definitions will be discussed in a later chapter.

This chapter is organised as follows. The first two sections further elaborate on the issue of prediction mismatch raised in Chapter 5. Prediction mismatch is expected in the setup of finite horizon MPC algorithms such as GPC and DMC and prevents straightforward stability proofs. Section 6.3 then shows how the move to infinite horizon reduces the prediction mismatch and hence facilitates a stability result. Practical means (based in dual mode ideas [83]) of deploying infinite prediction horizons are then given in the last two sections.

---

[*]They are sufficient, not necessary.

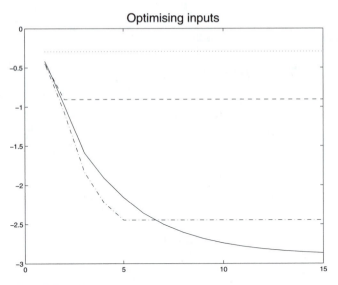

FIGURE 6.1

Optimised predicted input trajectories.

Summary: Stability can be guaranteed by the appropriate use of infinite hroizons and inclusion of the tail.

6.1   Prediction mismatch in MPC

One of the major weaknesses of the original predictive control algorithms such as DMC, GPC, IDCOM is the prediction structure used in the optimisation of performance. In simple terms the class of predictions over which the optimisation is performed is not necessarily closely matched to the resulting closed-loop responses. As such the optimisation could be ill posed; that is, finding a minimum for the given objective need not imply good performance. This weakness will be further illustrated using some examples.

6.1.1   Illustration of ill-posed objective

Take the example

$$y(z) = z^{-1}\frac{1 - 2z^{-1}}{(1 + z^{-1} + 0.9z^{-2})}u(z) \tag{6.1}$$

with tuning parameters $n_y = 15$, $n_u = 1, 2, 5, \lambda = 1$.

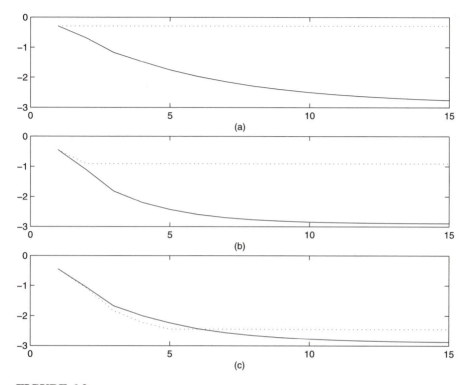

FIGURE 6.2

Optimised input predictions and closed-loop input trajectories.

Assuming no advance knowledge of set point changes, find the input trajectory optimising the objective (4.6) at the first sample when the set point change occurs. Plot
in Figure 6.1 the optimum predicted input trajectories using dotted, dashed, dash-dot
and solid lines, respectively, for $n_u = 1$, 2, 5 and a *global optimum*. It is clear that the
predicted input trajectories are not at all close to the *optimum* (especially for $n_u = 1$),
although, by chance, the first move may be fairly close.

Secondly, Figure 6.2 compares the open-loop predictions (dotted lines) to the corresponding closed-loop behaviour (solid lines). If the two are not similar then the optimisation must be ill posed. The open-loop optimised input trajectories are plotted
alongside the closed-loop responses that arise from the respective receding horizon
implementations. The plots are given in Figure 6.2a, b, c for $n_u = 1, 2, 5$ respectively.
Again it is clear that for small n_u there is a large difference between the optimal
open-loop predictions and the actual closed-loop behaviour.

Summary: The minimisation of finite horizon GPC may be ill posed; that is, for
small n_u the optimal prediction may bear only a small relationship to the actual or
desired behaviour.

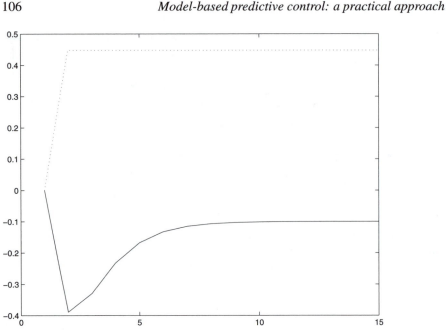

FIGURE 6.3

Optimum open- and closed-loop input trajectories.

6.1.2   Example where prediction mismatch causes instability

One can see the consequences of mismatch very clearly if one takes the example
of a nonminimum phase process and a low output horizon. In this case, because of
the low output horizon, the optimised inputs from minimising J can actually be the
opposite of the true optimal values.

Consider the example

$$y(z) = z^{-1} \frac{1 - 2z^{-1}}{1 - 0.9z^{-1}} \qquad (6.2)$$

with $n_y = 2$, $n_u = 1$, $\lambda = 1$. In Figure 6.3 is given the optimised open-loop predictions (at the step change) in dotted lines along with the global optimal closed-loop
trajectories in a solid line. The two sets of lines are clearly very different; in fact the
supposed *optimal* open-loop prediction for the input trajectory has the wrong sign
for this choice of tuning parameters. Unsurprisingly in this case the GPC control law
is destabilising.

This illustration is partly analogous to not looking far enough ahead when driving at
speed. Using the insights of Chapter 5 it is reasonable to expect that a higher value
of n_y, and preferably also n_u, is required to give good performance.

> **Summary:** Prediction mismatch can cause instability, even for stable open-loop
> processes.

6.1.3 Summary

It is clear from these examples that with finite horizons, n_y and n_u, the optimal open-loop predictions may not be well matched either to the consequent closed-loop behaviour arising from a receding horizon implementation or to the global optimum. In this case one may ask:

$$\boxed{\text{Is minimising } J \text{ meaningful?}}$$

The answer to this question is complex but was partially given in the previous chapter on tuning for GPC type algorithms. If one ensures that both n_u and n_y are sufficiently large, then the mismatch between open-loop predictions and closed-loop behaviour will be small and hence the minimisation will be well posed. What we mean by sufficiently large will become clear later in this chapter. It should be noted however, that computational limitations may imply that n_u (or the number of d.o.f. in the optimisation) must be small. In this case use of the GPC paradigm could inevitably give significant mismatch between the open-loop predictions and desired closed-loop behaviour and this must be taken into account in the design stage.

Summary: GPC algorithms can give poor performance, even for the nominal case.

- With n_u (and/or n_y) small, the GPC algorithm may have a poorly posed objective.

- The parameterisation of the d.o.f. may be poor.

- The algorithm may be set up with significant prediction mismatch.

Remark 6.1 *As ever the observations of this section also apply to MIMO processes but the effects will be harder to pinpoint due to the interactions taking place. However the same general guideline follows: choose the output horizon large enough and the input horizon as big as you can.*

6.2 Feedforward design in MPC

One of the supposed advantages of MPC is that it can make systematic use of advance knowledge of set points and disturbances. However, as will be shown here, one must treat this claim with caution [109]. Following the same lines as in Section 6.1, it will be shown here that the default choice of feedforward compensator (i.e. P_r of eqn.(4.26)) is often poor. This is also due to the mismatch between the assumptions made on the open-loop predictions and the actual closed-loop behaviour.

6.2.1   Structure of the set point prefilter

The default prefilter P_r is anticausual; that is, it contains powers of z rather than z^{-1}. Hence the current control action depends upon future set points. Assume that the default prefilter is given as

$$P_r = p_1 z + p_2 z^2 + \cdots + p_{n_y} z^{n_y} \qquad (6.3)$$

If one wanted to reduce the advance information used by the control law to say n_r steps, this is equivalent to assuming that $r_{k+n_r+i} = r_{k+n_r}$, $i > 0$. One can achieve this most simply by rewriting the prefilter as

$$P_r = p_1 z + p_2 z^2 + \cdots + p_{n_r-1} z^{n_r-1} + [p_{n_r} + p_{n_r+1} + \cdots + p_{n_y}] z^{n_r} \qquad (6.4)$$

If no advance knowledge is available, then $P_r = N_k(1)$, i.e. a simple gain.

6.2.2   Mismatch between predictions and actual behaviour

In the performance index minimisation, only a few control moves are allowed even though in the closed-loop it is known that a control move is ultimately to be allowed at every sample. Because the optimisation is working with just a few moves, it will optimise tracking over the whole output horizon, assuming just those few input changes. If there is a set point change towards the latter end of the horizon, the optimisation will request some movement from the inputs now in order to prepare for that set point change. That is, it starts moving too soon because the current minimisation does not have the freedom to use later control moves. One will then get a slow creeping of the output towards the expected set point with a faster movement once the set point has actually occurred.

One can illustrate this with the following example:

$$y = \frac{z^{-1}}{1 - 0.8 z^{-1}} u; \quad n_y = 15, \ n_u = 1 \text{ or } 3 \text{ or } 5, \ \lambda = 1 \qquad (6.5)$$

The closed-loop step responses are given in Figures 6.4, 6.5, 6.6 for $n_u = 1$, 3, 5 respectively where plots for a full 15th order prefilter P_r are given in dashed lines and for a 5th order prefilter in dotted lines. The set point is in solid lines.

It is clear that in each case one actually gets better performance by using less information about the future set point (a lower order prefilter), that is the dotted line is far better than the dashed line. The significance is less as n_u increases.

The cause of the poor performance is evident from an inspection of the input plots. If the order of the prefilter is too high, the input starts moving far too soon because in

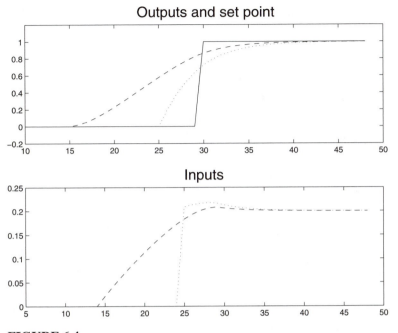

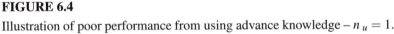

FIGURE 6.4

Illustration of poor performance from using advance knowledge – $n_u = 1$.

the prediction stage it sees no other option. An optimal solution to this problem, that is, how much advance information to use and how to design a better feedforward P_r, is an open question and very process dependent [109]. However, the designer should be aware of the dangers so that if necessary simple modifications, such as reducing the order of the prefilter, can be made.

> **Summary:** If the output horizon n_y is large and n_u is small, then you may get benefits from reducing the order of any feedforwards in the control law. Usually the effect is less noticeable for higher n_u.

6.2.3 Making better use of advance information

We have identified an apparent contradiction. That is, better control arises by ignoring available information. Even if we know that the set-point is to change in 10 samples time, we deny that information to the controller. This seems perverse in that surely an optimal controller law should make optimal use of the information and hence give better performance with than without that information.

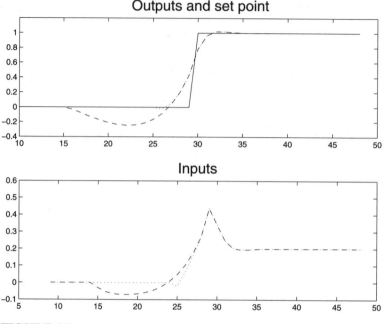

FIGURE 6.5

Illustration of poor performance from using advance knowledge – $n_u = 3$.

6.2.3.1 Understanding the misuse of future information

Of course the reason for this puzzle, already identified, is the mismatch between the predictions used in the cost optimisation and the actual closed-loop responses. During the optimisation, the algorithm is only aware of n_u control moves with which to optimise performance over a horizon n_y; in the closed-loop there will actually be n_y control moves. Therefore it gives a solution which is optimum assuming only n_u control moves but nonsense if more control moves are actually available. For instance, the algorithm sees a set point change 15 samples away but only has one or two control moves now to counter it when it would be better to wait until just a few samples before the set point change.

Summary: If there is a significant mismatch between the structure of the predictions and the actual closed-loop behaviour, then the optimisation may well be meaningless and at best ill posed.

6.2.3.2 Alternative choices for the feedforward

Unfortunately there is no easy solution to the problem identified here. The easiest solution is to increase n_u to match the amount of future information available;

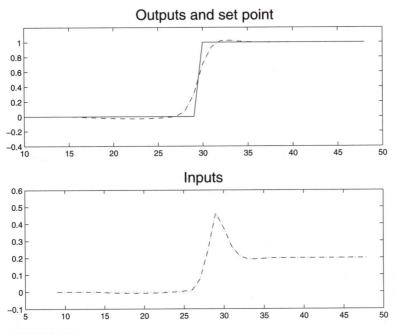

FIGURE 6.6

Illustration of poor performance from using advance knowledge – $n_u = 5$.

then the control algorithm has available control moves near the time of the set point changes. This of course is an unacceptable solution, as high values of n_u imply both high computational loads and can also give ill-conditioned optimisations.

A second and simpler solution is to limit the amount of set point information given to the control algorithm to be of the order of n_u and at most a bit less than the system closed-loop rise time. This improves matters and is a practical solution that is perhaps the only realistic option during constraint handling.

A third solution [109] is to use a different structure for the future control trajectory $\Delta\underset{\rightarrow}{\mathbf{u}}$, for instance, not assuming that all the control moves in the prediction are consequent but instead allowing gaps. Hence there is some freedom to deal with information appertaining to horizons further into the future. This discussion, however, is implicitly a part of the next few sections on stability and is left until then.

There is also an 'optimal' solution for the constraint free case [109]. It is known that the nominal closed-loop transferences, e.g. (4.35) are:

$$\begin{aligned}
\mathbf{u} &= \hat{D}[D_k(z)\Delta\hat{D} + N_k(z)\hat{N}]^{-1}P_r(z)\mathbf{r} \\
\mathbf{y} &= \hat{N}[D_k(z)\Delta\hat{D} + N_k(z)\hat{N}]^{-1}P_r(z)\mathbf{r}
\end{aligned} \tag{6.6}$$

These are clearly linear in the feedforward compensator P_r. Hence one can substitute closed-loop predictions (6.6) into a cost function of the form (4.6) and minimise w.r.t. the parameters of P_r. Hence this will therefore give optimal closed-loop use of feedforward. The GPC algorithm is then a two-stage design: (i) use (4.25) to define the loop compensator N_k, D_k and (ii) optimise closed-loop tracking by selecting P_r to minimise J w.r.t. closed-loop predictions. We omit the details of these computations as: (i) the result depends upon the choice of $\mathbf{r}$ and (ii) the most common use of GPC is for constraint handling. Thus it is doubtful that practitioners would ever use this procedure.

As yet there has not been a good summary in the literature of how to optimise the use of advance information for the constraint handling case. The author believes that practically future information is rarely available and even if it is, the information is probably not used in the way suggested possible by early papers [21].

Summary:

1. If you have information about future set point changes (or measurable disturbances) that enter the control law through feedforward compensators, then it is often wise to limit the amount of advance information used and hence avoid pitfalls such as those illustrated in Figures 6.4 and 6.5.

2. If you get poor performance from a GPC implementation on the nominal model, then there is likely to be a large mismatch between the predictions used in the cost function and the desired or actual closed-loop responses; this implies an ill-posed optimisation. If you want good performance, the prediction class should include a member close to the closed-loop optimum [138].

6.3   Infinite horizons imply stability

The observations of the previous sections were not fully understood at first and it was considered, at least in the academic literature, that the issues of both tuning and stability were major problems with GPC and variants. Although often in practice good control and good robustness margins were achieved, rigorous proofs were scarce. Many papers were written on heuristic guidelines for tuning parameters such as the control and output horizon and the effects of changing the control weighting. However, even given such guidelines, it was still only a posteriori stability checks that could be performed; that is, find the control law and then compute the implied closed-loop poles in the nominal case (unconstrained) – say from (4.36).

A more generic solution to this frustrating issue became common knowledge in the early 1990s (e.g. [22, 60, 87, 94, 107, 110, 138]) (though actually in the literature from the late 1970s, e.g. [55, 69]). This solution was to make use of the well known linear quadratic (LQ) optimal control results of the 1960's. However, before this result can be presented, the reader should first be familiar with one element necessary for the proof, that is *the tail*. The tail is defined in the next section and this is then followed by the proof itself.

6.3.1 Definition of *the tail*

It was stated in the introduction to this chapter that the inclusion of *the tail* within the class of predictions was important to giving a well-posed MPC algorithm; here a definition for the tail is given.

Let the optimal predictions at sampling instant k be given by the pair:

$$\Delta \underset{\rightarrow k-1}{\mathbf{u}} = \begin{bmatrix} \Delta \mathbf{u}_k \\ \Delta \mathbf{u}_{k+1|k}^T \\ \Delta \mathbf{u}_{k+2|k}^T \\ \vdots \end{bmatrix} ; \quad \underset{\rightarrow k}{\mathbf{y}} = \begin{bmatrix} \mathbf{y}_{k+1|k} \\ \mathbf{y}_{k+2|k}^T \\ \mathbf{y}_{k+3|k}^T \\ \vdots \end{bmatrix} \tag{6.7}$$

At the next sampling instant, that is $k+1$, the first components of this pair have already occurred and hence can no longer be called predictions. The part that still constitutes a prediction is called the tail, i.e.

$$\Delta \underset{\rightarrow k-1,\text{tail}}{\mathbf{u}} = \begin{bmatrix} \Delta \mathbf{u}_{k+1|k} \\ \Delta \mathbf{u}_{k+2|k}^T \\ \Delta \mathbf{u}_{k+3|k}^T \\ \vdots \end{bmatrix} ; \quad \underset{\rightarrow k,\text{tail}}{\mathbf{y}} = \begin{bmatrix} \mathbf{y}_{k+2|k} \\ \mathbf{y}_{k+3|k}^T \\ \mathbf{y}_{k+4|k}^T \\ \vdots \end{bmatrix} \tag{6.8}$$

It is important to note that the predictions given in the tail at $k+1$ were those computed at the previous sampling instant k.

A convenient stability proof is facilitated if the tail, that is the pair $(\Delta \underset{\rightarrow k,tail}{\mathbf{u}}, \underset{\rightarrow k,\text{tail}}{\mathbf{y}})$, are included in the class of possible predictions at $k+1$. That is the d.o.f. must be parameterised such that, for the nominal case, one can enforce:

$$\Delta \underset{\rightarrow k}{\mathbf{u}} = \Delta \underset{\rightarrow k-1,\text{tail}}{\mathbf{u}}; \quad \underset{\rightarrow k+1}{\mathbf{y}} = \underset{\rightarrow k,\text{tail}}{\mathbf{y}} \tag{6.9}$$

In more detail this implies, for example, that one could choose

$$\Delta \underset{\rightarrow k}{\mathbf{u}} = \begin{bmatrix} \Delta \mathbf{u}_{k+1} \\ \Delta \mathbf{u}_{k+2|k+1}^T \\ \Delta \mathbf{u}_{k+3|k+1}^T \\ \vdots \end{bmatrix} = \begin{bmatrix} \Delta \mathbf{u}_{k+1|k} \\ \Delta \mathbf{u}_{k+2|k}^T \\ \Delta \mathbf{u}_{k+3|k}^T \\ \vdots \end{bmatrix} \tag{6.10}$$

Summary: The tail is those parts of the predictions made at the previous sample which have still to take place. These should ideally be part of the current prediction class.

6.3.2 Infinite horizons and the tail

The reader is reminded that the arguments given next apply to the nominal case only. It is assumed that if a law is stabilising for the nominal case with reasonably fast convergence, then the gain and phase margins are likely to be large enough to imply a good degree of robustness. A detailed discussion of how to handle uncertainty in a systematic fashion is left for a later chapter.

Consider the infinite horizon cost[†] and optimisation

$$J_k = \min_{\Delta \mathbf{u}_{k+i}, \ i=0,1,\dots} \quad J = \sum_{i=1}^{\infty} \|\mathbf{r}_{k+i} - \mathbf{y}_{k+i}\|_2^2 + \lambda \|\Delta \mathbf{u}_{k+i-1}\|_2^2 \qquad (6.11)$$

After optimisation a whole input trajectory is defined, let this trajectory be given as

$$\Delta \mathbf{\underset{\rightarrow}{u}}_{k-1}^T = [\Delta \mathbf{u}_k^T, \Delta \mathbf{u}_{k+1|k}^T, \dots] \qquad (6.12)$$

Of this trajectory, the first element $\Delta \mathbf{u}_k$ is implemented. Now consider the optimisation to be performed at the next sampling instant $k+1$. Clearly, when dealing with the nominal case and selecting the input increments for all future time, the optimal values cannot differ from those computed previously[‡]. Hence

$$\begin{aligned} \Delta \mathbf{\underset{\rightarrow}{u}}_k^T &= [\ \Delta \mathbf{u}_{k+1}^T, \quad \Delta \mathbf{u}_{k+2|k+1}^T, \quad \dots \ ] \\ &= [\ \Delta \mathbf{u}_{k+1|k}^T, \quad \Delta \mathbf{u}_{k+2|k}^T, \quad \dots \ ] \end{aligned} \qquad (6.13)$$

That is the new optimum must coincide with the tail, $(\Delta \mathbf{\underset{\rightarrow}{u}}_k = \Delta \mathbf{\underset{\rightarrow}{u}}_{k-1,tail})$.

Now consider the implications on the cost function J using the notation that J_k is defined as the minimum of J at the kth sampling instant. From consideration of (6.11) it is clear that

$$J_{k+1} = J_k - \|\mathbf{r}_{k+1} - \mathbf{y}_{k+1}\|_2^2 - \lambda \|\Delta \mathbf{u}_k\|_2^2 \qquad (6.14)$$

That is J_k and J_{k+1} share common terms apart from the values associated to sample k which do not appear in J_{k+1}. Hence it is clear from (6.14) that

$$J_{k+1} \leq J_k \qquad (6.15)$$

[†]For simplicity the weights here are scalar but clearly arguments transfer easily to matrix weights. Also the results will apply to any weights which are monotonically increasing with prediction horizon.

[‡]This is a well-known observation in optimal control theory.

That is, the optimal predicted cost J_k can never increase. Moreover it is monotonically decreasing. This can be demonstrated by a simple contradiction. If $J_k = J_{k+1}$, then $\|\mathbf{r}_{k+1} - \mathbf{y}_{k+1}\|_2^2 + \lambda\|\Delta\mathbf{u}_k\|_2^2 = 0$ which implies the output is at set point and no input increment was required. This can only happen repeatedly if the plant is in steady state at the desired set point. If $J_k \neq J_{k+1}$, then $J_{k+1} < J_k$; that is there is a decrease in cost. In summary, if one uses $n_y = n_u = \infty$, then the cost function J becomes a Lyapunov function.

What is most significant here is that the stability proof does not entail computation of implied closed-loop poles (e.g. 4.36). That is, one can state in advance of computing the control law that it will be stabilising. This is called an a priori guarantee.

> **Summary:** Using J as a potential Lyapunov function has now become an accepted method for establishing a priori stability of MPC control laws.
>
> 1. Use of infinite horizons guarantees that J is Lyapunov.
>
> 2. Implicit in the proof that J is Lyapunov is the incorporation of the tail into the class of possible predictions.

6.3.3 Only the output horizon needs to be infinite

The above conclusions require only infinite output horizons because even if n_u is finite, the same arguments will follow. Consider the following optimisation

$$J_k = \min_{\Delta\mathbf{u}_{k+i}, \ i=0,1,\dots} \quad J = \sum_{i=1}^{\infty} \|\mathbf{r}_{k+i} - \mathbf{y}_{k+i}\|_2^2 + \sum_{i=1}^{n_u} \lambda\|\Delta\mathbf{u}_{k+i-1}\|_2^2 \tag{6.16}$$

Now enforce condition (6.13) which implies that $\Delta\mathbf{u}_{k+n_u|k+1} = 0$. Hence

$$J_{k+1} = J_k - \|\mathbf{r}_{k+1} - \mathbf{y}_{k+1}\|_2^2 - \lambda\|\Delta\mathbf{u}_k\|_2^2 \tag{6.17}$$

More importantly, at sampling instant $k+1$, one has freedom to select $\underset{\rightarrow}{\Delta\mathbf{u}}$ such that (6.13) is not satisfied if this makes J_{k+1} smaller still.

Remark 6.2 *It is worth noting that if n_u is small, although one can obtain a guarantee of stability there may still be significant prediction mismatch (as highlighted in Section 5.2.4.2) and this could result in poor transient performance.*

Remark 6.3 *Although it may not be obvious, the proofs given are tacitly assuming that $\lim_{i \to \infty} |\mathbf{r}_{k+i} - \mathbf{y}_{k+i}| = 0$ (or in the worst case is bounded). If not, the stability proof breaks down. Such an assumption may not be automatic where n_u is small, especially in the presence of large step changes or large disturbances. However, these issues do not belong here but rather in a discussion on feasibility (see Chapter 8).*

Summary:

- Infinite output horizons imply that the cost function is Lyapunov in the nominal case, which implies closed-loop stability.

- This result relies on the d.o.f. being such that condition (6.13) can always be satisfied (inclusion of the tail).

6.4 Stability proofs with constraints

The significance of the above a priori result is the constraint handling case. It was noted earlier that one could not easily assess stability during constraint handling, as linear analysis does not apply; the implied control law is nonlinear in the presence of constraints. However, the Lyapunov stability proof above also applies to the nonlinear case. The only requirement is that the constrained optimisation is always feasible; that is, there always exists a future input trajectory $\Delta \underrightarrow{\mathbf{u}}$ such that constraints are predicted to be satisfied over the entire future. This requirement is a demanding one that may not be met in practice; however, it will not concern us in this chapter.

Summary: The Lyapunov stability proof also applies during constraint handling.

6.4.1 Are infinite horizons impractical?

Although the solution of the infinite horizon optimal control problem implied by (6.11) is well known from the 1960's and straightforward to solve, this only applies to the unconstrained linear case. Practical MPC algorithms deal with constraints so that the actual optimisation (see Section 4.8.2) required takes the form:

$$J_k = \min_{\underrightarrow{\Delta \mathbf{u}}_{k-1}} J \quad \text{s.t.} \quad C\underrightarrow{\Delta \mathbf{u}}_{k-1} - \mathbf{d}_k \leq 0 \qquad (6.18)$$

where $C\underrightarrow{\Delta \mathbf{u}}_{k-1} - \mathbf{d}_k \leq 0$ represent the constraints ($\mathbf{d}_k$ is time varying; that is, it depends on the current state). This is a quadratic programming problem and hence can be solved in principle. However, one will immediately notice that:

- $\underrightarrow{\Delta \mathbf{u}}_{k-1}$ may be infinite dimensional.

- Matrix C and vector $\mathbf{d}_k$ have an infinite row dimension.

- As posed, the optimisation maybe intractable.

- In fact the only exception, tractable problem, is if both n_u is finite (which limits the number of d.o.f.) and there are only input constraints (no state/output constraints) so that $C, \mathbf{d}$ have finite row dimension.

Summary: In the presence of constraints a simplistic implementation of MPC based on large (infinite) horizons may be intractable.

6.4.2 Alternatives to optimal control

The remainder of this chapter will show how researchers have reformulated optimisation (6.18) to give a tractable problem with a useful solution.

It is interesting that historical developments in MPC started from simple controllers such as minimum variance control in the 1970s; and gradually increased complexity, for instance GPC in the 1980s; and then in the 1990s has returned to a form of optimal control, though in a more tractable formulation. The quest for the new millenium is perhaps to maintain the benefits of optimal control but increase transparency, to reduce computational load and extensions to more demanding cases such as model uncertainty and nonlinearity.

This increase in the complexity of the control strategy reflects the increase in computing power available and hence problems such as (6.18) look increasingly tractable whereas in the 1970s they were not and hence complexity in itself does not need to be unreasonable in some scenarios.

6.5 Dual mode control – an overview

The terminology dual mode control [83] is more often associated with nonlinear control strategies and is not strictly valid for the strategies to be dicussed here. However, it conveys the principle philosophy, hence its adoption in this book. The most popular realisations of infinite horizon MPC have a dual mode form and so the next section gives an overview of the principal components in a dual mode strategy.

6.5.1 What is dual mode control?

This is a control strategy which has two modes. One mode is used when the system is far away from steady state or far from the operating point. The second mode is used when close to the desired operating point. Hence there is an implied switching between one mode of operation and another as the process converges to the desired state.

In MPC, the notation dual mode does not imply a switching between modes in real time, but rather is a description of how the predictions are set up. Hence, consider a prediction over n_y steps; then one could take the first n_c steps to be within mode 1 and the remaining steps to be in mode 2.

$$\underset{\rightarrow}{\mathbf{x}} = [\underbrace{\mathbf{x}_{k+1|k}, \mathbf{x}_{k+2|k}, \dots, \mathbf{x}_{k+n_c|k}}_{\text{Mode 1}}, \underbrace{\mathbf{x}_{k+n_c+1|k}, \dots, \mathbf{x}_{k+n_y|k}}_{\text{Mode 2}}] \qquad (6.19)$$

One must emphasise that this switching is in the predictions only. The closed-loop control law has a single mode but uses dual mode predictions in the optimisation. Moreover it would be usual to take n_y in mode 2 to be infinite.

6.5.2 The structure of dual mode predictions

The most usual assumption, at least in MPC, is that the first n_c control moves $\Delta\mathbf{u}_{k+i}$, $i = 0, \dots, n_c - 1$, are free and that the remaining moves $\Delta\mathbf{u}_{k+i}$, $i \geq n_c$ are given by a fixed feedback law. So for instance, in the state-space case the predictions could be given by

$$\begin{aligned} \mathbf{x}_{k+i} &= A\mathbf{x}_{k+i-1} + B\mathbf{u}_{k+i-1}, & \mathbf{u}_{k+i-1} \text{ are d.o.f.} & \quad i = 1, 2, \dots, n_c \\ \mathbf{x}_{k+i} &= [A - BK]\mathbf{x}_{k+i-1}, & \mathbf{u}_{k+i-1} = -K\mathbf{x}_{k+i-1}, & \quad i > n_c \end{aligned} \qquad (6.20)$$

Efficient realisations of this will be discussed in a different chapter. Also modification of this concept for other model forms and offset free prediction is omitted as straightforward (see Chapters 3 and 4).

Summary: Dual mode control refers to the predictions being separated into near transients (mode 1 behaviour) and asymptotic predictions (mode 2 behaviour). It is normal for mode 2 behaviour to be given by a known control law.

6.5.3 Overview of MPC dual mode algorithms

Dual mode describes a philosophy and hence many different variants can be developed. This section illustrates just one such variant and several others will be discussed in the following chapter. Assume a state-space model for the following [§]. **Preliminaries:**

1. Define a control law, say $\mathbf{u} = -K\mathbf{x}$.

2. Define a terminal invariant (see Section 11.7) region $\mathbb{S}$ in the phase plane for state $\mathbf{x}$, assuming the given feedback. This region may be ellipsoidal or polyhedral. Assume $\mathbb{S}$ is set up so that given recursive use of the nominal feedback $\mathbf{u} = -K\mathbf{x}, \mathbf{x} \in \mathbb{S}$ implies constraints are satisfied.

[§]However, one can equally apply this principle to transfer function models.

3. Define n_c and compute the prediction equations as discussed in (6.20).

4. Define an on-line performance measure as

$$J_k = \sum_{i=1}^{\infty} \|\mathbf{r}_{k+i} - \mathbf{y}_{k+i}\|_2^2 + \lambda \|\Delta\mathbf{u}_{k+i-1}\|_2^2 \tag{6.21}$$

and the vector of d.o.f.

$$\Delta\underset{\rightarrow k-1}{\mathbf{u}} = [\Delta\mathbf{u}_k^T, \ldots, \Delta\mathbf{u}_{k+n_c-1}^T]^T \tag{6.22}$$

5. Define the inequalities $C\Delta\underset{\rightarrow k-1}{\mathbf{u}} - \mathbf{d}_k \leq 0$ as ensuring constraint satisfaction during the first n_c steps only.

6.5.4 Two possible dual mode algorithms

Algorithm 6.1 Dual mode control 1: *At each sampling instant, minimise the infinite horizon cost as follows;*

$$\min_{\Delta\underset{\rightarrow k}{\mathbf{u}}} \quad J_k \quad \text{s.t.} \quad \begin{cases} C\Delta\underset{\rightarrow k}{\mathbf{u}} - \mathbf{d}_k \leq 0 \\ \mathbf{u}_{k+i} = -K\mathbf{x}_{k+i}, \ i \geq n_c \end{cases} \tag{6.23}$$

Algorithm 6.2 Dual mode control 2: *Choose a state feedback K and find a set S satisfying (11.6). Then minimise the infinite horizon cost as follows;*

$$\min_{\Delta\underset{\rightarrow k}{\mathbf{u}}} \quad J_k \quad \text{s.t.} \quad \begin{cases} C\Delta\underset{\rightarrow k}{\mathbf{u}} - \mathbf{d}_k \leq 0 \\ \mathbf{u}_{k+i} = -K\mathbf{x}_{t+i}, \ i > n_c \\ \mathbf{x}_{k+n_c+1|k} \in S \end{cases} \tag{6.24}$$

Remark 6.4 *The difference in these two algorithms is the constraint $\mathbf{x}_{k+n_c+1|k} \in S$ in the latter algorithm. This constraint ensures that the predictions do not violate constraints beyond n_c. However, it is not always applied explicitly but sometimes is taken for granted as in the former algorithm. The relevance of this will be discussed in Chapter 8.*

6.5.5 Is a dual mode strategy guaranteed stabilising?

It was stated throughout this chapter that the inclusion of certain components is sufficient to guarantee nominal stability. The main two of these are the use of infinite horizons and the inclusion of the tail. These two together are sufficient to prove that the cost function J is a Lyapunov function. Dual mode control uses infinite horizons; hence it is only required to demonstrate the inclusion of the tail.

Assume that the terminal feedback is K so that the closed-loop dynamic in the far future is given by $\mathbf{x}_{k+i+1} = \Phi\mathbf{x}_{k+i}$, $\Phi = A - BK$. Compare the predictions at two consequent sampling instants:

$$\Delta\underset{\rightarrow k-1}{\mathbf{u}} = \begin{bmatrix} \Delta\mathbf{u}_k \\ \Delta\mathbf{u}_{k+1|k} \\ \vdots \\ \Delta\mathbf{u}_{k+n_u-1|k} \\ -K\mathbf{x}_{k+n_u|k} \\ -K\Phi\mathbf{x}_{k+n_u|k} \\ -K\Phi^2\mathbf{x}_{k+n_u|k} \\ \vdots \end{bmatrix} \qquad : \qquad \begin{bmatrix} \Delta\mathbf{u}_{k+1} \\ \vdots \\ \Delta\mathbf{u}_{k+n_u-1|k+1} \\ \Delta\mathbf{u}_{k+n_u|k+1} \\ -K\mathbf{x}_{k+n_u+1|k+1} \\ -K\Phi\mathbf{x}_{k+n_u+1|k+1} \\ \vdots \end{bmatrix} = \Delta\underset{\rightarrow k}{\mathbf{u}} \qquad (6.25)$$

In order for these two vectors to be the same one needs to ensure

$$\begin{aligned} \Delta\mathbf{u}_{k+i|k+1} &= \Delta\mathbf{u}_{k+i|k}, \qquad i = 1, ..., n_u - 1 \\ \Delta\mathbf{u}_{k+n_u|k+1} &= -K\mathbf{x}_{k+n_u|k} \\ \mathbf{x}_{k+n_u+1|k+1} &= \Phi\mathbf{x}_{k+n_u|k} \end{aligned} \qquad (6.26)$$

Clearly the first of these is straightforward, as $\Delta\mathbf{u}_{k+i|k+1}$, $i = 1, ..., n_u - 1$ constitute d.o.f. Second, one can force $\Delta\mathbf{u}_{k+n_u|k+1} = -K\mathbf{x}_{k+n_u|k}$, as this also is a d.o.f. Finally, the satisfaction of the first two implies that $\mathbf{x}_{k+n_u+1|k+1} = \Phi\mathbf{x}_{k+n_u|k}$ as

$$\Delta\mathbf{u}_{k+n_u|k+1} = -K\mathbf{x}_{k+n_u|k} \quad \Rightarrow \quad \mathbf{x}_{k+n_u+1|k+1} = \Phi\mathbf{x}_{k+n_u|k} \qquad (6.27)$$

> **Summary:** The tail is in the class of possible dual mode predictions. Hence dual mode predictions lend themselves to an MPC law with guaranteed stability.

6.5.6 How do dual mode predictions make infinite horizon MPC more tractable ?

The main motivation for using dual mode predictions is that they give a handle on the predictions over an infinite horizon. That is, the part of the prediction in mode 2 can be analysed using standard linear analysis, as they are given by the implementation of simple linear feedback law. Hence although the predictions evolve over an infinite horizon, one can define these predictions with just a finite number of d.o.f. and moreover it can be shown that for many practical cases, the terminal set S (see algorithm 6.2) is finitely determined [33]; hence the implied optimisation is finite dimensional.

> **Summary:** Dual mode predictions allow a reduction in the number of d.o.f. and constraints to be handled while still allowing the use of infinite input and output prediction horizons.

6.6 Implementation of dual mode MPC

The previous section illustrated, in principle, that the dual mode concept gives guaranteed stability when deployed in MPC. It remains now to give the details of how the control law is computed and in particular how the dual mode concept facilitates the minimisation of an infinite horizon cost function and the handling of constraints over an infinite horizon.

The key ingredient in dual mode strategies is the separation: mode 1 is totally free whereas mode 2 is predetermined. This separation limits the number of d.o.f. to those in mode 1 hence giving tractable optimisations. Second, by predetermining the behaviour of mode 2 predictions, one opens them up to linear analysis. In particular, it is noted that the predictions in mode 2 are deterministic given the predicted value of the state $\mathbf{x}_{k+n_c}$ which has an affine dependence on the d.o.f. $\Delta \underset{\rightarrow k}{\mathbf{u}}$ and hence is easy to handle.

This section will be split into four parts. The first shows how to compute a quadratic cost function over an infinite horizon with linear predictions. The second shows how to construct a performance index with dual mode predictions and the third shows how to set up constraint equations for dual mode predictions. These parts are then united to define the dual mode MPC algorithm.

6.6.1 The cost function for linear predictions over infinite horizons

Assume that predictions are deterministic, then one can evaluate the corresponding infinite horizon cost function using a Lyapunov type of equation. For instance, assume

$$\mathbf{x}_{k+i+1} = \Phi \mathbf{x}_{k+i} = \Phi^i \mathbf{x}_k; \quad \mathbf{u}_{k+i} = -K\mathbf{x}_{k+i} = -K\Phi^i \mathbf{x}_k \tag{6.28}$$

and that

$$J_k = \sum_{i=0}^{\infty} \mathbf{x}_{k+i+1}^T Q \mathbf{x}_{k+i+1} + \mathbf{u}_{k+i}^T R \mathbf{u}_{k+i} \tag{6.29}$$

Substitute into (6.29) from (6.28):

$$
\begin{aligned}
J &= \sum_{i=0}^{\infty} \mathbf{x}_k^T (\Phi^{i+1})^T Q \Phi^{i+1} \mathbf{x}_k + \mathbf{x}_k^T K^T (\Phi^i)^T R \Phi^i K \mathbf{x}_k \\
&= \sum_{i=0}^{\infty} \mathbf{x}_k^T \underbrace{[(\Phi^{i+1})^T Q \Phi^{i+1} + K^T (\Phi^i)^T R \Phi^i K]}_{P} \mathbf{x}_k \\
&= \mathbf{x}_k^T P \mathbf{x}_k
\end{aligned} \tag{6.30}
$$

where

$$P = \sum_{i=0}^{\infty} (\Phi^{i+1})^T Q \Phi^{i+1} + (\Phi^i)^T K^T R K \Phi^i$$

It can be shown be simple substitution that one can solve for P using a Lyapunov equation

$$\Phi^T P \Phi = P - \Phi^T Q \Phi - K^T R K \qquad (6.31)$$

Summary: For linear predictions (6.28), the quadratic cost function (6.29) takes the form below where P is determined from a Lyapunov equation.

$$J = \mathbf{x}_k^T P \mathbf{x}_k \qquad (6.32)$$

6.6.2 Forming the cost function for dual mode predictions

The state evolves according to model (6.20). Hence it is convenient to separate cost (6.29) into two parts for modes 1 and 2.

$$J = J_1 + J_2; \quad \begin{array}{l} J_1 = \sum_{i=0}^{n_c-1} \mathbf{x}_{k+i+1}^T Q \mathbf{x}_{k+i+1} + \mathbf{u}_{k+i}^T R \mathbf{u}_{k+i} \\ J_2 = \sum_{i=0}^{\infty} \mathbf{x}_{k+n_c+i+1}^T Q \mathbf{x}_{k+n_c+i+1} + \mathbf{u}_{k+n_c+i}^T R \mathbf{u}_{k+n_c+i} \end{array} \qquad (6.33)$$

The cost J_1 can be constructed using the usual arguments of Section 3.2. For instance

$$\underset{\rightarrow}{\mathbf{x}} = P_{xx} \mathbf{x}_k + H_x \underset{\rightarrow}{\mathbf{u}}_{k-1} \qquad (6.34)$$

where $\underset{\rightarrow}{\mathbf{x}}, \underset{\rightarrow}{\mathbf{u}}$ are defined for a horizon n_c. Hence

$$J_1 = [P_{xx} \mathbf{x}_k + H_x \underset{\rightarrow}{\mathbf{u}}_{k-1}]^T \text{diag}(Q)[P_{xx} \mathbf{x}_k + H_x \underset{\rightarrow}{\mathbf{u}}_{k-1}] + \underset{\rightarrow}{\mathbf{u}}_{k-1} \text{diag}(R) \underset{\rightarrow}{\mathbf{u}}_{k-1} \qquad (6.35)$$

Using the result of the previous section, it is clear that the cost J_2 depends only on $\mathbf{x}_{k+n_c|k}$ and can be represented as

$$J_2 = \mathbf{x}_{k+n_c|k}^T P \mathbf{x}_{k+n_c|k}; \quad \Phi^T P \Phi = P - \Phi^T Q \Phi - K^T R K \qquad (6.36)$$

One can use the last block rows of prediction (6.34) to find a prediction for $\mathbf{x}_{k+n_c|k}$; define this as

$$\mathbf{x}_{k+n_c|k} = P_{n_c} \mathbf{x}_k + H_{n_c} \underset{\rightarrow}{\mathbf{u}}_{k-1} \qquad (6.37)$$

where P_{n_c}, H_{n_c} are the n_c^{th} block rows of P_{xx}, H_x respectively. Hence

$$J_2 = [P_{n_c} \mathbf{x}_k + H_{n_c} \underset{\rightarrow}{\mathbf{u}}_{k-1}] P [P_{n_c} \mathbf{x}_k + H_{n_c} \underset{\rightarrow}{\mathbf{u}}_{k-1}] \qquad (6.38)$$

Finally one can combine J_1, J_2 from (6.35, 6.38) to give:

$$\begin{array}{l} J = [P_{xx} \mathbf{x}_k + H_x \underset{\rightarrow}{\mathbf{u}}_{k-1}]^T \text{diag}(Q)[P_{xx} \mathbf{x}_k + H_x \underset{\rightarrow}{\mathbf{u}}_{k-1}] + \underset{\rightarrow}{\mathbf{u}}_{k-1} \text{diag}(R) \underset{\rightarrow}{\mathbf{u}}_{k-1} \\ + [P_{n_c} \mathbf{x}_k + H_{n_c} \underset{\rightarrow}{\mathbf{u}}_{k-1}] P [P_{n_c} \mathbf{x}_k + H_{n_c} \underset{\rightarrow}{\mathbf{u}}_{k-1}] \end{array} \qquad (6.39)$$

This can be simplified and is summarised next.

Summary: The dual mode performance index takes a simple quadratic form with n_c block d.o.f.

$$J = \underset{\rightarrow k-1}{\mathbf{u}}^T S \underset{\rightarrow k-1}{\mathbf{u}} + \underset{\rightarrow k-1}{\mathbf{u}}^T L \mathbf{x}_k + k \tag{6.40}$$

where $S = H_x^T \text{diag}(Q) H_x + \text{diag}(R) + H_{n_c}^T P H_{n_c}$, $L = 2[H_x^T \text{diag} Q P_{xx} + H_{n_c}^T P P_{n_c}]$ and k does not depend on the d.o.f. $\underset{\rightarrow k-1}{\mathbf{u}}$.

6.6.3 Constraint handling with dual mode predictions

It will be demonstrated in Section 11.7 [33] that for an autonomous model, such as given in (6.28), one could construct a maximal admissible set S_{max} which was such that for any state within the set, the evolution of inputs and states, under state feedback $\mathbf{u} = -K\mathbf{x}$, would be such that no constraints are violated. Let S_{max} be represented by linear inequalities:

$$S_{max} = \{\mathbf{x} : C_{max}\mathbf{x} - \mathbf{d}_{max} \leq 0\} \tag{6.41}$$

In dual mode predictions, the behaviour is equivalent to the autonomous model (6.28) for $i \geq n_c$; hence constraints are guaranteed to be satisfied during mode 2 if and only if $\mathbf{x}_{k+n_c|k} \in S_{max}$. One can substitute for $\mathbf{x}_{k+n_c|k}$ from eqn.(6.37) to give

$$C_{max}[P_{n_c}\mathbf{x}_k + H_{n_c}\underset{\rightarrow k-1}{\mathbf{u}}] - \mathbf{d}_{max} \leq 0 \tag{6.42}$$

Constraints during mode 1 can be tested (see eqn. 4.90) explicitly using expressions of the form

$$C\underset{\rightarrow}{\mathbf{u}} - \mathbf{d} \leq 0 \tag{6.43}$$

Remark 6.5 *It is common practice in the literature to use the terminology terminal set or target set for S_{max}; that is the set in which $\mathbf{x}$ must lie after n_c steps. As will be apparent later the definition of this set (and implicitly of K) is a key design parameter in dual mode MPC.*

Summary: Dual mode predictions satisfy constraints over the entire horizon if and only if

$$C\underset{\rightarrow}{\mathbf{u}} - \mathbf{d} \leq 0 \quad \text{and} \quad C_{max}[P_{n_c}\mathbf{x}_k + H_{n_c}\underset{\rightarrow k-1}{\mathbf{u}}] - \mathbf{d}_{max} \leq 0 \tag{6.44}$$

6.6.4 Computing the dual mode MPC control law

We now have all the components required to define a dual mode MPC algorithm. The cost is given in (6.40) and the constraints are given by (6.44). Hence the on-line computation[¶] is:

$$\min_{\underset{\rightarrow}{\mathbf{u}}} \quad \mathbf{u}_{\underset{\rightarrow k-1}{}}^T S \mathbf{u}_{\underset{\rightarrow k-1}{}} + \mathbf{u}_{\underset{\rightarrow k-1}{}}^T L \mathbf{x}_k \quad \text{s.t.} \quad \text{constraints (6.44)} \tag{6.45}$$

It is noted that one no longer needs to make explicit the constraint $\mathbf{u}_{k+n_c+i|k} = -K\mathbf{x}_{k+n_c+i-1|k}$, as this is implicit in the definition of J and the definition of the constraints.

Remark 6.6 *Recall that* $\underset{\rightarrow}{\mathbf{u}}_{k-1}$ *in the optimisation comprises the first* n_c *control moves only. The choice of* n_c *therefore controls the number of d.o.f. in the optimisation.*

Summary: Despite the use of infinite horizons, a dual mode implementation of MPC reduces to a quadratic program with a finite, and possibly small, number of d.o.f. and a finite number of constraints.

6.6.5   Remarks on stability and performance of dual mode control

A remarkable result has arisen. *Irrespective of the choice of K* (as long as it is stabilising), the dual mode MPC algorithm is guaranteed stable for the nominal case. In essence, fixing the d.o.f. beyond some point in the horizon does not affect the stability proof so long as any prediction assumptions are reproducible at subsequent sampling instants, that is, inclusion of the tail. As will be seen in Chapter 7 this still allows a lot of flexibility in the design.

The main result of this chapter is based on the incorporation of the two key components of an infinite horizon and the tail. However as noted in the introduction there are two other equally important components which affect performance: (i) prediction mismatch and (ii) feasibility.

- If the choice of terminal control law K is poor, then there could still be significant prediction mismatch; and when n_c is small, this could lead to poor performance.

- If the choice of K is highly tuned such as to allow good performance, then the terminal set S_{max} may be small and the optimisation of (6.45) may be infeasible; that is, it may not be possible to satisfy all the constraints. The particular problem here is the requirement that the state enter the terminal set S_{max} in at most n_c steps and this could be problematic where n_c is small.

[¶]Clearly this is for state feedback actually implemented as in (4.52). The reader will need to rework some of the details for other prediction models.

Feasibility now becomes a big issue, as the constraints are not just on the inputs, which can always be satisfied, but also on the states. State constraints may be inconsistent with input constraints and there may be no systematic way of correcting this due to the artificial nature of the terminal constraint $\mathbf{x}_{k+n_c|k} \in S_{max}$. In the event of infeasibility the algorithm is undefined.

Subsequent chapters will look at how the choice of terminal control law affects both performance and feasibility.

Summary:

1. The use of infinite horizons and incorporation of the tail allows stability guarantees.

2. The dual mode paradigm gives an implementation of infinite horizon MPC which is computationally tractable and relatively straightforward to implement.

3. However, dual mode MPC may not have either a good performance or a large region of applicability. A good compromise between large terminal sets and good performance is a significant design issue which is tackled in the next chapter.

7

Closed-loop paradigm

The previous chapter has shown that dual mode predictions facilitate the design of MPC algorithms with both guaranteed stability and the potential for good performance. This chapter will consider some consequent themes:

1. What is a good way to set up a dual mode MPC algorithm?

2. What are some common choices of dual mode MPC algorithms?

3. Are dual mode algorithms necessary?

First the reader will be formally introduced to the closed-loop paradigm which is a numerically robust and insightful way of implementing dual mode algorithms. This paradigm is then used to develop the insights given in the remainder of the chapter.

Summary: The closed-loop paradigm is an alternative mechanism for implementing MPC algorithms which can have advantages.

7.1 Introduction to the closed-loop paradigm

The closed-loop paradigm (CLP) was originally proposed as part of an algorithm stable generalised predictive control (SGPC,[60]) but can be abbreviated hereafter as stable predictive control (SPC) to take account of other strategies. Increasingly this paradigm is being adopted by researchers in MPC due to the good properties it introduces into the predictive control problem. For instance:

1. It gives better numerical conditioning [104, 113] of the optimisation which is essential for open-loop unstable plant.

2. It gives useful insight into the structure of dual mode strategies in that it shows how far one is away from optimum due to constraint handling.

3. It makes robustness analysis more straightforward (e.g. [40, 66, 114]) even for the constrained case.

It can be shown that algebraically, the CLP is identical to an equivalent open-loop paradigm (OLP) strategy (all the work in this book so far is OLP as it uses open-loop predictions). Hence ultimately whether one uses a CLP or an OLP approach will depend a little on personal preference and the particular scenario.

7.1.1   Overview of the CLP concept

The basic idea is to choose a stabilising control law (which could be arbitrary but usually is chosen with some aim) and assume that this law is present throughout the predictions. Alternatively one could take the view that the control law is hardwired into the prediction computation which implies one has pseudo closed-loop predictions. Philosophically one could consider this as analogous to the control law used in mode 2 of dual mode MPC with the only difference being that the terminal law is now also deployed during mode 1. The prediction equations in Section 7.1.2 will make this clearer.

7.1.1.1   Trival guarantee of stability

If the predictions are set up as being based on a stabilising control law and hence based on a stable closed loop, then one has a potentially trivial stability proof; simply use the control trajectories associated to this underlying law, which by definition is selected to be stabilising. However, this strategy is unlikely to be optimal unless the underlying control law itself is optimal, which need not be the case in general. Moreover, the closed-loop predictions associated to this underlying law may not satisfy constraints and hence could be infeasible; this invalidates any stability guarantee outside the associated maximal admissible set (MAS) [33].

Finally of course the reader would realise that so far such a strategy has no direct link to MPC, as it reduces to implementing a known stabilising control law with no reference to how this law may be derived. A partial link could be made by assuming that the underlying stabilising control law is derived via an MPC algorithm.

7.1.1.2   Modification of CLP predictions to allow constraint handling

The two major contributions of MPC are that it gives a systematic design of a stabilising control law for the MIMO case in the constraint free case and it allows systematic on-line constraint handling; that is, it can propose valid (satisfying constraints) input trajectories even when the state is outside the MAS of the implied underlying linear control law. The use of CLP predictions would only be beneficial if both of these advantages could be retained. The first is automatic, using MPC to design the unconstrained control law. The incorporation of the second is explained next.

In dual mode control constraint handling is achieved by choosing control moves in mode 1 (these are the d.o.f.) so that the state in mode 2 is inside the MAS S_{max}

associated to the control law implied in mode 2. With the CLP the same philosophy is adopted in that d.o.f. are introduced in mode 1 and these are used to ensure that the mode 2 predictions are inside the MAS S_{max} of the underlying control law.

The only remaining question is, how are d.o.f. introduced into the mode 1 predictions when mode 1 is already based on a stabilising control law? The answer is simple; the d.o.f. are simply perturbations to the control moves determined by the underlying law. These perturbations are selected to ensure constraint satisfaction during the mode 1 predictions and to ensure that the mode 2 predictions are inside the associated MAS.

The beauty of the CLP approach is that the d.o.f. can be set up as perturbations about the most desirable performance [137], that is, the performance that arises from unconstrained control. Hence one gains insight into the impact of constraints by viewing the magnitude of these perturbations. That is, the larger the perturbations, the further the constrained optimal behaviour is away from unconstrained behaviour and hence the greater the impact constraints are having on performance.

Stability reduces to ensuring that the perturbations converge to zero which is intuitively easy to interpret. Even more significantly, the desirable control law is explicitly included in the predictions which gives a good structure to the predictions and improves numerical conditioning. Moreover it alleviates problems with prediction mismatch, as by definition the class of predictions can include the desirable closed-loop behaviour.

Summary: The closed-loop paradigm uses perturbations to the unconstrained optimal control law as d.o.f. This gives good insight into the impact of constraints on performance and improves the conditionning of the optimisation.

7.1.1.3 Illustration of prediction structure with the CLP

It will be easier to understand the previous two sections after seeing the prediction structure so this is illustrated next. The OLP (as given in Chapter 3) and CLP prediction structures are given together to aid comparison.

$$
\underset{\rightarrow OLP}{\mathbf{u}} = \begin{bmatrix} \mathbf{u}_{k|k} \\ \mathbf{u}_{k+1|k} \\ \vdots \\ \mathbf{u}_{k+n_c-1|k} \\ \hline -K\mathbf{x}_{k+n_c|k} \\ \vdots \\ -K\Phi^{n_y-n_c}\mathbf{x}_{k+n_c|k} \end{bmatrix} ; \quad \underset{\rightarrow CLP}{\mathbf{u}} = \begin{bmatrix} -K\mathbf{x}_k + \mathbf{c}_k \\ -K\mathbf{x}_{k+1|k} + \mathbf{c}_{k+1} \\ \vdots \\ -K\mathbf{x}_{k+n_c|k} + \mathbf{c}_{k+n_c-1} \\ -K\mathbf{x}_{k+n_c|k} \\ \vdots \\ -K\Phi^{n_y-n_c}\mathbf{x}_{k+n_c|k} \end{bmatrix} \tag{7.1}
$$

From here it is clear that the d.o.f. in the OLP predictions are the first n_c control moves; that is $\mathbf{u}_{k+i}, i = 0, ..., n_c - 1$, whereas in the CLP predictions the d.o.f. are the perturbations $\mathbf{c}_{k+i}, i = 0, ..., n_c - 1$. The mode 2 predictions are the same, that is, based on $\mathbf{x}_{k+n_c|k}$.

Theorem 7.1 *The OLP and CLP paradigms give an identical prediction class.*

Proof: This is obvious, as during mode 1 the corresponding terms are

$$\mathbf{u}_{k+i|k} \quad : \quad K\mathbf{x}_{k+i|k} + \mathbf{c}_{k+i} \qquad (7.2)$$

Clearly $\mathbf{u}_{k+i|k}$, $\mathbf{c}_{k+i}$ have the same dimension and hence can be selected to ensure equivalence. The mode 2 predictions are the same if $\mathbf{x}_{k+n_c|k}$ is the same which is true if the mode 1 predictions are the same. □

Corollary 7.1 *The OLP and CLP paradigms parameterise the d.o.f. in a different way.*

Proof: This is obvious from the predictions. The significance will be seen later in the formulation of the cost function J using both prediction sets; in general the CLP predictions give a better conditioned optimisation. □

Summary: The OLP and CLP are equivalent in the space covered but have different parameterisations of the d.o.f. In fact the CLP can be viewed as a means of normalising the space of optimisation variables which improves conditioning.

7.1.2 CLP predictions

For completeness this section gives the CLP predictions (analogous to those derived in Sections 3.2, 3.3 for OLP) which can be deployed in MPC algorithms.

7.1.2.1 CLP predictions for state-space models

For simplicity the details of how to incorporate integral action are omitted. These follow the same lines as given in Section 2.3.4.

The equations within the prestabilised loop (during prediction) are

$$\mathbf{x}_{k+i|k} = A\mathbf{x}_{k+i-1|k} + B\mathbf{u}_{k+i}; \quad \mathbf{u}_{k+i} = -K\mathbf{x}_{k+i|k} + \mathbf{c}_{k+i} \qquad (7.3)$$

Removing the dependent variable $\mathbf{u}_{k+i}$ one gets:

$$\mathbf{x}_{k+i|k} = [A - BK]\mathbf{x}_{k+i-1|k} + B\mathbf{c}_{k+i}; \quad \mathbf{u}_{k+i} = -K\mathbf{x}_{k+i|k} + \mathbf{c}_{k+i} \qquad (7.4)$$

Simulating these forward in time with $\Phi = A - BK$ one gets:

$$\underline{\mathbf{x}}_k = \underbrace{\begin{bmatrix} \Phi \\ \Phi^2 \\ \Phi^3 \\ \vdots \end{bmatrix}}_{P_{cl}} \mathbf{x}_k + \underbrace{\begin{bmatrix} B & 0 & 0 & \dots \\ \Phi B & B & 0 & \dots \\ \Phi^2 B & \Phi B & B & \dots \\ \vdots & \vdots & \vdots & \vdots \end{bmatrix}}_{H_c} \underline{\mathbf{c}}_k \qquad (7.5)$$

or in more compact form

$$\underset{\rightarrow k}{\mathbf{x}} = P_{cl}\mathbf{x}_k + H_c \underset{\rightarrow k}{\mathbf{c}} \tag{7.6}$$

The corresponding input predictions can be written as

$$\underset{\rightarrow k}{\mathbf{u}} = \underbrace{\begin{bmatrix} -K \\ -K\Phi \\ -K\Phi^2 \\ \vdots \end{bmatrix}}_{P_{clu}} \mathbf{x}_k + \underbrace{\begin{bmatrix} B & 0 & 0 & \ldots \\ -KB & B & 0 & \ldots \\ -K\Phi B & -KB & B & \ldots \\ \vdots & \vdots & \vdots & \vdots \end{bmatrix}}_{H_{cu}} \underset{\rightarrow k}{\mathbf{c}} \tag{7.7}$$

or

$$\underset{\rightarrow k}{\mathbf{u}} = P_{clu}\mathbf{x}_k + H_{cu} \underset{\rightarrow k}{\mathbf{c}} \tag{7.8}$$

The state after n_c steps will be denoted as

$$\mathbf{x}_{k+n_c|k} = P_{cl2}\mathbf{x}_k + H_{c2} \underset{\rightarrow k}{\mathbf{c}} \tag{7.9}$$

Summary: Predictions are affine in the current state and the d.o.f. and hence have an equivalent form to (3.1).

7.1.2.2 CLP predictions with transfer function models

In this case (see Chapter 3 to remind yourself of notation) the nominal control law (eqn.(4.25)) takes the form:

$$\Delta \mathbf{u}_k = P_r \underset{\rightarrow}{\mathbf{r}} - N_k \underset{\leftarrow}{\mathbf{y}} - \check{D}_k \Delta \underset{\leftarrow}{\mathbf{u}} \tag{7.10}$$

One can then perturb the implied control action by a simple modification as follows:

$$\Delta \mathbf{u}_k = P_r \underset{\rightarrow}{\mathbf{r}} - N_k \underset{\leftarrow}{\mathbf{y}} - \check{D}_k \Delta \underset{\leftarrow}{\mathbf{u}} + \mathbf{c}_k \tag{7.11}$$

or with the more usual representation of eqn.(4.31)

$$D_k(z)\Delta \mathbf{u}_k = P_r(z)\mathbf{r}_{k+1} - N_k(z)\mathbf{y}_k + \mathbf{c}_k \tag{7.12}$$

The model equation is

$$A(z)\mathbf{y} = b(z)\Delta \mathbf{u} \tag{7.13}$$

and hence closed-loop predictions can be formulated by solving the model and controller simultaneously. This is illustrated using the Toeplitz/Hankel methodology given in Chapter 3.

1. The model and controller equations over the prediction horizon are:

$$\begin{aligned} C_A \underset{\rightarrow}{\mathbf{y}} + H_A \underset{\leftarrow}{\mathbf{y}} &= C_{zb}\Delta \underset{\rightarrow}{\mathbf{u}} + H_{zb}\Delta \underset{\leftarrow}{\mathbf{u}} \\ C_{z^{-1}N_k} \underset{\rightarrow}{\mathbf{y}} + H_{z^{-1}N_k} \underset{\leftarrow}{\mathbf{y}} &= -C_{D_k}\Delta \underset{\rightarrow}{\mathbf{u}} - H_{D_k}\Delta \underset{\leftarrow}{\mathbf{u}} - \underset{\rightarrow}{\mathbf{c}} \end{aligned} \tag{7.14}$$

2. Eliminate $\underset{\rightarrow}{\Delta\mathbf{u}}$ using the second of these equations:

$$\underset{\rightarrow}{\Delta\mathbf{u}} = C_{D_k}^{-1}[C_{z^{-1}N_k}\underset{\rightarrow}{\mathbf{y}} + H_{z^{-1}N_k}\underset{\leftarrow}{\mathbf{y}} - H_{D_k}\underset{\leftarrow}{\Delta\mathbf{u}} - \underset{\rightarrow}{\mathbf{c}}]$$

3. Then substitute $\underset{\rightarrow}{\Delta\mathbf{u}}$ into the first equation

$$
\begin{aligned}
C_A\underset{\rightarrow}{\mathbf{y}} + H_A\underset{\leftarrow}{\mathbf{y}} &= C_{zb}C_{D_k}^{-1}[C_{z^{-1}N_k}\underset{\rightarrow}{\mathbf{y}} + H_{z^{-1}N_k}\underset{\leftarrow}{\mathbf{y}} - H_{D_k}\underset{\leftarrow}{\Delta\mathbf{u}} - \underset{\rightarrow}{\mathbf{c}}] \\
&\quad + H_{zb}\underset{\leftarrow}{\Delta\mathbf{u}} \\
[C_A - C_{zb}C_{D_k}^{-1}C_{z^{-1}N_k}]\underset{\rightarrow}{\mathbf{y}} &= [C_{zb}C_{D_k}^{-1}H_{z^{-1}N_k} + H_A]\underset{\leftarrow}{\mathbf{y}} + [C_{zb}C_{D_k}^{-1}H_{D_k} + H_{zb}]\underset{\leftarrow}{\Delta\mathbf{u}} \\
&\quad - C_{zb}C_{D_k}^{-1}\underset{\rightarrow}{\mathbf{c}} \\
\underbrace{[C_{D_k}C_A - C_{zb}C_{z^{-1}N_k}]}_{C_{P_c}}\underset{\rightarrow}{\mathbf{y}} &= [C_{zb}H_{z^{-1}N_k} + C_{D_k}H_A]\underset{\leftarrow}{\mathbf{y}} + [C_{zb}H_{D_k} + C_{D_k}H_{zb}]\underset{\leftarrow}{\Delta\mathbf{u}} \\
&\quad - C_{zb}\underset{\rightarrow}{\mathbf{c}} \\
\underset{\rightarrow}{\mathbf{y}} &= C_{P_c}^{-1}\{[C_{zb}H_{z^{-1}N_k} + C_{D_k}H_A]\underset{\leftarrow}{\mathbf{y}} \\
&\quad + [C_{zb}H_{D_k} + C_{D_k}H_{zb}]\underset{\leftarrow}{\Delta\mathbf{u}} - C_{zb}\underset{\rightarrow}{\mathbf{c}}\}
\end{aligned}
$$

(7.15)

Note that P_c is the implied closed-loop polynomial derived with the given control law.

4. In a similar way one can derive that

$$\underset{\rightarrow}{\Delta\mathbf{u}} = C_{P_c}^{-1}[\underbrace{[C_{zb}H_{z^{-1}N_k} + C_{D_k}H_A]}_{P_y}\underset{\leftarrow}{\mathbf{y}} + \underbrace{[C_{zb}H_{D_k} + C_{D_k}H_{zb}]}_{P_u}\underset{\leftarrow}{\Delta\mathbf{u}} - C_A\underset{\rightarrow}{\mathbf{c}}]$$ (7.16)

Remark 7.1 *The reader will notice in (7.15, 7.16) a nice separation between the part of the predictions dependent on past information and that dependent on the perturbations $\underset{\rightarrow}{\mathbf{c}}$. This neat separation aids both insight and computation. For instance if the underlying control law is 'optimal', then one would ideally choose $\underset{\rightarrow}{\mathbf{c}} = 0$.*

Summary: CLP predictions take the same neat form as given in (3.1) with the exception that the d.o.f. are expressed in terms of the perturbation **c** rather than in terms of **u**. Hence these can be substituted into MPC in a straightforward manner.

7.1.3 CLP structure

It is useful to have a picture in one's head of how the CLP paradigm might be implemented. As there is an implied underlying control law throughout the prediction structure, this could be viewed as hardwired into the system; that is, one could actually implement this control law. The MPC algorithm then supplies the perturbations **c** to improve performance or for constraint handling. For the control law of (7.12) this scenario can be represented by Figure 7.1:

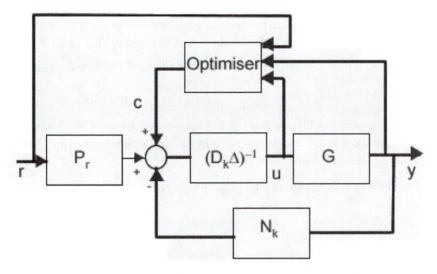

FIGURE 7.1

Control loop with the closed-loop paradigm.

Here the fixed control law has parameters P_r, N_k, D_k. This is considered as being always implemented. The past inputs and outputs are used to form the predictions (7.15, 7.16) hence these along with the future set point are used in the optimisation stage which produces the best choice of $\underrightarrow{c}$, the perturbation to the underlying control law.

A similar diagram could be constructed for state-space implementations.

Remark 7.2 *The informed reader will notice a strong link with reference governor strategies [34, 37, 38]. This is unsurprising, as reference governor strategies are often MPC algorithms that simply deploy an alternative parameterisation of the d.o.f. to simplify computation [119].*

7.2 Setting up an MPC problem with the closed-loop paradigm

Next the MPC implementation of the CLP is introduced in more detail. This follows the same pattern as for the OLP given in Chapters 3 and 4 and hence is shown here only briefly.

The main components of MPC are predictions, a performance index, d.o.f. and constraints. Hence there is a need to define each of these components for the CLP.

1. The d.o.f. are the perturbations $\mathbf{c}$ identified in Figure 7.1. It is usual to define these as:

$$\mathbf{c}_{\rightarrow k}^T = [\mathbf{c}_{k|k}^T, \dots, \mathbf{c}_{k+n_c-1|k}^T]; \quad \mathbf{c}_{k+n_c+i|k} = 0, \ i \geq 0 \qquad (7.17)$$

That is, assume a finite number* (n_c) of nonzero values for $\mathbf{c}$. Beyond n_c the perturbations are zero and the loop acts in a linear fashion and is equivalent to mode 2 of the dual mode predictions.

2. The performance index is the the usual infinite horizon cost

$$J_k = \min_{\mathbf{c}_{k+i}, \ i=0,1,\dots} \quad J = \sum_{i=1}^{\infty} \|\mathbf{r}_{k+i} - \mathbf{y}_{k+i}\|_2^2 + \lambda \|\Delta \mathbf{u}_{k+i-1}\|_2^2 \qquad (7.18)$$

J needs to be formulated in terms of the d.o.f. $\mathbf{c}_{\rightarrow}$.

3. The predictions (7.15, 7.16) come from simulating the loop of Figure 7.1, hence they are closed-loop predictions. There is a dependence on initial values in the loop and the assumptions on the perturbations $\mathbf{c}_{\rightarrow}$.

4. The constraints (4.90) or (6.44) need to be formulated in terms of their dependence on the predictions, e.g. (7.15, 7.16) or (7.6, 7.8, 7.9).

Summary: To design a CLP MPC law, one needs to define the usual components of predictions, constraints, and objective in terms of the d.o.f. $\mathbf{c}_{\rightarrow}$.

7.2.1 Setting up the cost function and computing the control law for state-space models

Assuming that $\mathbf{c}_{\rightarrow}$ comprises only a finite number of nonzero values, then the cost function can be set up as for dual mode predictions where mode 2 is the part where $\mathbf{c} = 0$. Hence, following the procedure of Section 6.6.2 and using prediction equations (7.6, 7.8, 7.9) give the cost function as

$$\begin{aligned}
J &= [P_{cl}\mathbf{x}_k + H_c \mathbf{c}_{\rightarrow k}]^T \mathrm{diag}(Q)[P_{cl}\mathbf{x}_k + H_c \mathbf{c}_{\rightarrow k}] \\
&\quad + [P_{clu}\mathbf{x}_k + H_{cu}\mathbf{c}_{\rightarrow k}]^T \mathrm{diag}(R)[P_{clu}\mathbf{x}_k + H_{cu}\mathbf{c}_{\rightarrow k}] \\
&\quad + [P_{cl2}\mathbf{x}_k + H_{c2}\mathbf{c}_{\rightarrow k}]^T P[P_{cl2}\mathbf{x}_k + H_{c2}\mathbf{c}_{\rightarrow k}] \\
&= \mathbf{c}_{\rightarrow}^T S_c \mathbf{c}_{\rightarrow} + 2\mathbf{c}_{\rightarrow}^T S_{cx}\mathbf{x} + k
\end{aligned} \qquad (7.19)$$

where

$$\begin{aligned}
S_c &= H_c^T \mathrm{diag}(Q)H_c + H_{cu}^T \mathrm{diag}(R)H_{cu} + H_{c2}^T P H_{c2} \\
S_{cx} &= H_c^T \mathrm{diag}(Q)P_{cl} + H_{cu}^T \mathrm{diag}(R)P_{clu} + H_{c2}^T P P_{cl2}
\end{aligned} \qquad (7.20)$$

---

*Hence this gives an equivalence to mode 1 of the dual mode predictions.

and k does not depend upon $\underset{\rightarrow}{c}$ and hence can be ignored. This is quadratic in the degrees of freedom $\underset{\rightarrow}{c}$.

The MPC law is derived by minimising w.r.t. to the d.o.f., that is

$$\min_{\underset{\rightarrow}{c}} J = \underset{\rightarrow}{c}^T S_c \underset{\rightarrow}{c} + 2\underset{\rightarrow}{c}^T S_{cx}\mathbf{x} \tag{7.21}$$

which implies that the optimal unconstrained $\underset{\rightarrow}{c}$ is given from

$$\underset{\rightarrow}{c} = -S_c^{-1} S_{cx}\mathbf{x} \tag{7.22}$$

Notably this is a state feedback; that is, the optimal unconstrained perturbation is given by an additional feedback loop. The underlying control law is defined as $\mathbf{u} = -K\mathbf{x} + \mathbf{c}$ where $\underset{\rightarrow}{c} = -S_c^{-1} S_{cx}\mathbf{x}$.

Summary: The CLP MPC control law in the unconstrained state-space case is equivalent to a fixed state feedback:

$$\mathbf{u} = -\underbrace{[K + S_c^{-1} S_{cx}]}_{K_c}\mathbf{x} \tag{7.23}$$

7.2.2 Including the constraints for state-space models

Take the constraint equations of (6.44) which apply to dual mode algorithms. One can simply substitute in the predictions of (7.6, 7.8, 7.9) as appropriate in order to express the constraints in terms of $\underset{\rightarrow}{c}$.

The constraints are represented by the inequalities

$$C\Delta\underset{\rightarrow}{\mathbf{u}} - \mathbf{d} \leq 0 \quad \text{and} \quad C_{max}\mathbf{x}_{k+n_c|k} - \mathbf{d}_{max} \leq 0 \tag{7.24}$$

Substituting these gives inequalities of the form

$$\begin{aligned} C[P_{clu}\mathbf{x}_k + H_{cu}\underset{\rightarrow k}{c}] - \mathbf{d} &\leq 0 \\ C_{max}[P_{cl2}\mathbf{x}_k + H_{c2}\underset{\rightarrow k}{c}] - \mathbf{d}_{max} &\leq 0 \end{aligned} \tag{7.25}$$

Hence the entire constraint set (for the state-space models) is

$$\underbrace{\begin{bmatrix} CH_{cu} \\ C_{max}H_{c2} \end{bmatrix}}_{C_{clp}} \underset{\rightarrow k}{c} + \underbrace{\begin{bmatrix} CP_{clu} \\ C_{max}H_{c2} \end{bmatrix}}_{P_{clp}} \mathbf{x}_k - \underbrace{\begin{bmatrix} \mathbf{d} \\ \mathbf{d}_{max} \end{bmatrix}}_{d_{clp}} \leq 0 \tag{7.26}$$

Remark 7.3 *For transfer function models one would use predictions (7.15,7.16) for mode 1 constraints, i.e.*

$$C[P_y \underset{\leftarrow}{y} + P_u \underset{\leftarrow}{u} + H_{cu}\underset{\rightarrow}{c}] - \mathbf{d} \leq 0 \tag{7.27}$$

The MAS (mode 2 constraints) would take the form

$$C_{maxtf} \begin{bmatrix} \underset{\leftarrow}{\Delta \mathbf{u}} \\ \underset{\leftarrow}{\mathbf{y}} \end{bmatrix} + C_{max} H_{c2} \underset{\rightarrow}{\mathbf{c}} - \mathbf{d}_{maxtf} \leq 0 \qquad (7.28)$$

where matrices $C_{maxtf}, C_{max}, \mathbf{d}_{maxtf}$ are easy to define.

Summary: The constraint equations are affine in the d.o.f. $\underset{\rightarrow}{\mathbf{c}}$, the current state and the limits.

7.2.3 The constrained optimisation

The MPC optimisation (state-space case) using the CLP is given as

$$\min_{\underset{\rightarrow}{\mathbf{c}}} \ J = \underset{\rightarrow}{\mathbf{c}}^T S_c \underset{\rightarrow}{\mathbf{c}} + 2 \underset{\rightarrow}{\mathbf{c}}^T S_{cx} \mathbf{x} \quad \text{s.t.} \ \ C_{clp} \underset{\rightarrow}{\mathbf{c}} - P_{clp} \mathbf{x} - \mathbf{d}_{clp} \leq 0 \qquad (7.29)$$

The constrained control law is then defined as

$$\mathbf{u}_k = -K\mathbf{x}_k + \mathbf{c}_k \qquad (7.30)$$

where $\mathbf{c}_k$ is the first element of the optimising $\underset{\rightarrow}{\mathbf{c}}$.

Remark 7.4 *1. A similar method will give the solution in the transfer function case where the control law would take the form of (7.12).*

2. As with the OLP, the matrices in (7.29) depend upon the weights and the number of d.o.f. n_c. However, additionally they also depend upon the choice of terminal control law.

Summary: The computation of the MPC control law using the CLP reduces to the solution of a quadratic programming problem in the d.o.f. $\underset{\rightarrow}{\mathbf{c}}$ where $\mathbf{c}$ is the perturbation to the underlying (or mode 2) control law.

7.3 Different choices for mode 2 of dual mode control

Dual mode control with the OLP and the closed-loop paradigm give the same result; they differ only in the parameterisation of the d.o.f. but give identical solutions in the absence of numerical errors. That is, the control law of (7.30) can be shown to be identical to that which arises from the corresponding OLP minimisation (6.45). This is because one has used the same cost function and equivalent d.o.f.

From here on this book will adopt the CLP where infinite horizon algorithms are used; this because, as will be illustrated, there are benefits to be had over the OLP implementation.

However, first in this section we will give an overview of the different forms of dual mode control that have been popularised in the literature, followed by some comments on their respective strengths and weaknesses. The reader is reminded of the terminology:

1. **Terminal control law:** Sometimes denoted terminal conditons, this is the implied control law used in mode 2 of the predictions.

2. **Terminal region or target set:** This is the maximal admissible set[†] associated to the terminal control law.

Two main tuning parameters affect dual mode control. The most obvious is the number of d.o.f. but the second and in practice equally important one is the selection of the terminal control law. The following sections look at the effects of changing the terminal control law.

Remark 7.5 *If n_c is large, say around the system settling time, then the choice of terminal control law will have negligible effect on the unconstrained control law (7.23), as all the transients will occur in mode 1 and mode 2 will comprise of only small numbers. In other words*

$$\lim_{n_c \to \infty} [K + S_c^{-1} S_{cx}] = K_{opt} \qquad (7.31)$$

where K_{opt} is the optimal controller minimising (6.11); this can be derived using optimal control algorithms.

However, it is usually assumed that for computational reasons n_c is small. In this case the choice of terminal control law has a significant effect as it shapes much of the transients which are now in mode 2 of the predictions.

> **Summary:** The choice of terminal control law has a significant impact on unconstrained and constrained performance for small n_c.

7.3.1 Dead beat terminal conditions (SGPC)

The dead beat choice was popularised in the MPC field by [22, 60, 87] but actually known earlier (e.g. [55, 69]). Historically these were the first insightful suggestions of how to guarantee, a priori, the stability of MPC algorithms.

[†]In nonlinear MPC and computational simpler algorithms this set may be chosen to be an alternative invariant set and hence possible smaller than the MAS.

The key idea was to use a dead beat form of terminal mode, i.e. force the predicted output to be identically equal to the set point for n steps (n the model order). Hence, for the nominal case, the predicted output would remain at the set point thereafter, assuming that the corresponding predicted $\Delta \mathbf{u}$ were also zero.

One can easily demonstrate the consistency of the above statements. Take the incremental form of the model difference equation, for instance eqn.(3.14)

$$y_{k+1} = -A_1 y_k \cdots - A_{n+1} y_{k-n} + b_1 \Delta u_k + \ldots + b_{n-1} \Delta u_{k-n+1}$$

Next set all the past outputs equal to a constant r and past input increments equal to zero. Now compute the next predicted output:

$$\begin{aligned} y_{k+1} &= -A_1 r \cdots - A_{n+1} r + b_1 \Delta u_k + \ldots + b_{n-1} \Delta u_{k-n+1} \\ &= -[A_1 + \cdots + A_{n+1}] r + 0 \\ &= r \end{aligned} \tag{7.32}$$

(Recall that $[A_1 + \cdots + A_{n+1}] = -1$ because $A(z) = a(z)\Delta(z)$.)
What remains now is to ask how the requirement that

$$y_{k+i|k} = r, \quad i = n_c + n, n_c + n + 1, \ldots, n_c + 2n \tag{7.33}$$

might be best ensured and implemented in a control law design?

7.3.1.1 Implementing a dead beat terminal constraint

The dual mode paradigm gives the most obvious insight into how this constraint can be enforced[‡]. It is clear that the terminal control law should be a dead beat control law as by definition this can ensure condition (7.33). Hence

- In the state-space case one should choose K such that

$$(A - BK)^n = 0 \tag{7.34}$$

  where n is the state dimension. Then, for instance

$$\mathbf{x}_{k+n_c+n|k} = (A - BK)^n \mathbf{x}_{k+n_c|k} = 0 \tag{7.35}$$

  A dead beat feedback can be constructed via a pole placement design where all the poles are on the origin.

- In the transfer function case one should choose N_k, D_k such that the implied closed-loop poles are zero, i.e.

$$P_c = AD_k + bN_k = 1 \tag{7.36}$$

[‡]The reader may like to note that some earlier papers [22, 87] used an alternative and numerically inferior [104] approach which, however, may be the only option for nonlinear systems.

It is straightforward to show from the closed-loop transferences of Figure 7.1 (or otherwise) that condition (7.36) implies that $y_{k+n_c+n|k} = r$ (assuming $P_r(1) = N_k(1)$). For simplicity it would be usual to take the minimum order N_k, D_k satisfying (7.36) hence giving a unique solution. Other choices are related to robust design covered in a later chapter.

Having selected the terminal control law all the other details follow automatically from the sections on dual mode control and the CLP.

Summary: A terminal condition given by a constant output and unchanging input can be implemented by selecting the mode 2 control law as a dead beat controller.

7.3.1.2 Strengths and weaknesses of dead beat terminal conditions

Although innovative when first proposed, actually dead beat terminal conditions are not to be recommended in general. There are some obvious weaknesses to the law K_c of for instance (7.23) when K is dead beat:

1. Dead beat control is known to use very active input signals. Hence the terminal region S_{max} will be small.

2. As S_{max} is small, the use of small n_c will give rise to frequent infeasibility. To improve the volume of feasibility regions one would need a large n_c with a consequent increase in the on-line computational load.

3. Dead beat control would tend to give poor performance, measured by way of typical quadratic performance indices such as (6.11). If n_c is small, a significant part of the transients are *under* dead beat control and hence performance is expected to be poor.

4. Dead beat control usually has poor robustness; some of this poor sensitivity will inevitably be inherited by the associated MPC law (e.g. (7.23)) when n_c is small.

In fact the only obvious positive to dead beat conditions is that the transients have a fixed length, that is $n + n_c$, and as such the constraint equations are also limited to this horizon. This means the number of inequalities implied, in for instance (7.26), may be far fewer than with other terminal control laws. This simplifies (reduces the complexity of) the optimisation.

Summary: Dead beat terminal conditions should be avoided unless, such as in some nonlinear applications, it is the only stabilising control law which can be easily computed.

7.3.2   No terminal control (NTC)

In fact this is same as the standard GPC/DMC algorithm but with an infinite output horizon as in eqn.(6.16). By no terminal control (NTC) it is intended that the terminal control law is $\Delta\mathbf{u}_{k+n_c+i|k} = 0$, $i \geq 0$ for the transfer function formulation or can be realised by a zero state feedback $K = 0$ (implemented as in eqn.(4.61)) in the state feedback case. This is the most popular approach deployed in industry, more especially because the standard DMC algorithm can be used. The reader will note that DMC and GPC are in fact dual mode strategies although such a link is not usually made. However, there are significant weaknesses in using NTC strategies which will be detailed next:

1. Mode 2 is given by open-loop behaviour. In many cases this may imply poor performance and hence one returns to the earlier arguments on prediction mismatch and the possible impact on closed-loop performance.

2. The terminal region may be large (infinite) if there are only input constraints because the terminal input is zero. However, conversely S_{max} may be small in the presence of state constraints, as the behaviour is uncontrolled in mode 2.

3. The combination of an infinite output horizon and <u>small n_c</u> will give open-loop behaviour, essentially equivalent to potentially poorly performing integral control (see Section 5.2.4.2). This because the cost function is dominated by output errors and the cost minimisation is effectively subject to the constraint that the asymptotic error is as small as possible with n_c control moves.

4. It is not appropriate to unstable open-loop systems, as the implied terminal cost (6.32) with open-loop behaviour is infinite.

> **Summary:** NTC is easy to code and has good feasibility in the absence of state constraints. However, where n_c is small it may result in unnecessarily detuned performance. The reader is reminded that if one can deploy large n_c, then all algorithms will perform well.

7.3.3   Terminal mode by elimination of unstable modes (EUM)

The main weakness of SGPC (dead beat terminal conditions) is that it can give over-tuned control whereas the weaknesses of NTC terminal conditions are that it is potentially undertuned and not applicable to unstable processes.

An equivalent of NTC for unstable processes was developed independently by several authors, some using the CLP approach [41, 42, 107, 110] with transfer functions and some using the OLP approach [94] with state-space models. The basic problem was how to eradicate the unstable predictions from mode 2 when the implied mode 2 control law was zero. The natural answer deployed by all the authors was to ensure that the predicted state at the end of mode 1 was solely inside the stable manifold so that the mode 2 predictions would then be convergent.

For convenience here we use the pseudonym EUM for eliminate unstable modes. EUM forms a middle ground between SGPC and NTC. SGPC places an explicit dead beat constraint on mode 2 predictions which is severe; EUM on the other hand deploys a milder but explicit dead beat constraint on states solely within the unstable manifold. NTC places no terminal constraint at all. For stable processes EUM and NTC are equivalent.

The EUM algorithm selects only the unstable modes to penalise in mode 2 and hence it is algebraically more difficult to set up than SGPC and NTC. This is because the dead beat constraint on the unstable manifold is not easily realised as a fixed state feedback and hence does not fit easily into the dual mode template of the previous chapter. Instead one has to include the EUM constraint explicity as a set of linear equalities to be satisfied.

One can still construct the dual mode predictions and in line with the rest of this chapter, here we will do this using a CLP as opposed to an OLP. We will illustrate the procedure for both state-space and transfer function models.

7.3.3.1 Predictions with cancellation of unstable modes: state-space models

Let a state-space matrix have some unstable eigenvalues. Decompose the system into stable and unstable modes using the eigenvalue/vector decomposition:

$$A = [W_s, W_u]\text{diag}[\Lambda_s, \Lambda_u] \begin{bmatrix} V_s^T \\ V_u^T \end{bmatrix} \qquad (7.37)$$

where subscript s is used for stable and u for unstable. Clearly if a state lies solely in the stable manifold of A, then it must satisfy:

$$V_u^T \mathbf{x} = 0 \qquad (7.38)$$

Given this, the predicted state evolution would follow

$$\mathbf{x}_{k+i|k} = A^i \mathbf{x}_k = W_s \Lambda_s^i V_s^T \mathbf{x}_k \qquad (7.39)$$

To be more specific, assuming that $V_u^T \mathbf{x}_{k+n_c|k} = 0$, then the mode 2 predictions are given by

$$\mathbf{x}_{k+n_c+i|k} = W_s \Lambda_s^i V_s^T \mathbf{x}_{k+n_c|k}; \quad \mathbf{u}_{k+n_c+i} = 0 \qquad (7.40)$$

Given predictions (7.40) it is easy (following the procedure of Section 6.6.1) to formulate the implied mode 2 cost as[§]:

$$\sum_{i=1}^{\infty} \mathbf{x}_{k+n_c+i|k}^T Q \mathbf{x}_{k+n_c+i|k} = \mathbf{x}_{k+n_c|k} P_{eum} \mathbf{x}_{k+n_c|k} \qquad (7.41)$$

[§]Recall that the implied control is zero and hence is not included.

7.3.3.2 EUM control law for the state-space case

The unconstrained EUM control law contains 3 components:

1. Mode 1 predictions

2. Mode 2 predictions

3. The constraint on $\mathbf{x}_{k+n_c|k}$

Assuming that the mode 1 predictions are given by (7.6, 7.8, 7.9) and the mode 2 cost is given by (7.41) then the optimisation required is:

$$\min_{\underset{\rightarrow}{\mathbf{c}}} \; J = \underset{\rightarrow}{\mathbf{c}}^T S_{eum} \underset{\rightarrow}{\mathbf{c}} + 2 \underset{\rightarrow}{\mathbf{c}}^T S_{eumx} \mathbf{x} \quad \text{s.t.} \quad \begin{cases} V_u^T H_{c2} \underset{\rightarrow}{\mathbf{c}} + V_u^T P_{cl2} \mathbf{x} = 0 \\ C[P_{clu} \mathbf{x}_k + H_{cu} \underset{\rightarrow}{\mathbf{c}}_k] - \mathbf{d} \le 0 \end{cases} \tag{7.42}$$

where

$$S_{eum} = H_c^T \mathrm{diag}(Q) H_c + H_{cu}^T \mathrm{diag}(R) H_{cu} + H_{c2}^T P_{eum} H_{c2}$$
$$S_{eumx} = H_c^T \mathrm{diag}(Q) P_{cl} + H_{cu}^T \mathrm{diag}(R) P_{clu} + H_{c2}^T P_{eum} P_{cl2}$$

Summary: The required optimisation for EUM MPC has both equality and inequality constraints but otherwise has a standard form.

7.3.3.3 EUM predictions for the transfer function case

In this case it is easier to represent the predictions using transfer functions. Consider the predictions of (3.19) and represent by an equivalent transfer function; this procedure was outlined in Section 3.7.2.1. Hence

$$\underset{\rightarrow}{y}(z) = \frac{p(z) + b(z) \underset{\rightarrow}{u}(z)}{A(z)} \tag{7.43}$$

where $p(z) = [1 \; z^{-1} \; z^{-2} \; \ldots](H_{zb} \underset{\leftarrow}{\Delta u} - H_A \underset{\leftarrow}{\Delta u})$. In order for the predictions to be stable there must be no unstable modes. Let $A(z) = a^+(z) a^-(z)$ where a^+ contains the unstable factors. Therefore, in order for the predictions to converge the following must be true:

$$p(z) + b(z) \underset{\rightarrow}{u}(z) = a^+ \phi \tag{7.44}$$

where ϕ is a finite order polynomial. Eqn. (7.44) places a constraint on the possible choices for $\underset{\rightarrow}{u}$; the result is given in eqn.(3.86).

Remark 7.6 *The approach to EUM given here does not make use of predictions (7.15). However, this is because the terminal control law is $K = 0$. The prestabilisation is performed algebrically through (7.44) and hence has some equivalence to the CLP philosophy [107, 113].*

7.3.3.4 Strengths and weaknesses of EUM

These are analogous to the strengths and weaknesses of NTC but with the obvious improvement of being applicable to unstable open-loop plant. In fact for stable processes NTC and EUM are identical[¶].

The main reservation over EUM is similar to that for SGPC. If n_c is small then the constraint (7.38) may be difficult to realise; that is, the associated MAS for $K = 0$ and **x** in the stable manifold could be small. This limitation could be significant and is due to the restriction that the input moves be zero beyond n_c, a weakness which could also hamper performance due to the arguments on prediction mismatch. Examples of the poor performance due to the use of EUM constraints are given in [132, 133].

Remark 7.7 *Other work (e.g. [107]) extended EUM and looked at the minimal constraints that one could impose on the predictions to ensure convergence and so maximise the d.o.f. for performance. The key observation is that the terminal control moves need not be zero and this relaxes the condition that the state be in the stable manifold in only n_c steps. However, although interesting, this is largely superceded by later work and hence is discussed no further here.*

> **Summary:** The EUM control law is slightly more involved than other dual mode algorithms due to the need to include an equality constraint (i.e. 7.38). However, this can be incorporated quite neatly if desired (see eqn.(3.86)) hence MPC still reduces to a straightforward QP optimisation.

7.3.4 Terminal mode is optimal (LQMPC)

One logical choice [113, 137, 145] for K is in fact that which minimises the infinite horizon cost in the constraint free case. Define such an algorihm as linear quadratic optimal MPC (LQMPC). The advantage of such a choice is that for any n_c, if the unconstrained optimum is feasible, then the dual mode algorithm will find that solution; *that is no prediction mismatch in the unconstrained case.* Moreover, if n_c is large enough, then the dual mode algorithm can find the optimal for the constrained infinite dimensional optimisation. This is because, at least in the prediction stage when one expects the outputs to settle, there always exists an n_c large enough so that the optimum $x_{k+n_c|k} \in S_{max}$ for any given stabilising K. In practice this n_c may not be very large.

No details are given for this algorithm, as they are implicit in Section 7.1.1. It should be noted that in the constraint free case this algorithm will given the optimal $\underset{\rightarrow}{c} = 0$; that is, the term in (7.22) $\underset{\rightarrow}{c} = -S_c^{-1} S_{cx} = 0$, which implies that $S_{cx} = 0$ and the cost

---

[¶]Of course the EUM paradigm does have the flexibility to allow poles near the stability boundary to be treated as unstable so that they do not play a part in mode 2 predictions.

is centered on the origin, i.e. $J = \underset{\rightarrow}{\mathbf{c}}^T S_c \underset{\rightarrow}{\mathbf{c}}$! In this case the size of the perturbations $\mathbf{c}$ is a direct measure of the distance from the unconstrained optimum.

However it is still necessary to consider strengths and weaknesses.

1. The major weakness is that for a well-tuned K the terminal region S_{max} may be small. Hence for a large region of attraction one would need a large n_c.

2. The major strength is that the prediction class includes the unconstrained optimum and hence there is a good chance that one will have no prediction mismatch which implies good closed-loop performance.

3. The predictions evolve over an infinite horizon, so irrespective of n_c, the constraint equations may contain a large number of rows. Hence it is less straightforward to use structure in the constraint equations to simplify the optimisation, even when there are no state constraints.

Summary: Using a mode 2 control law of K_{opt} ensures a well posed optimisation and a well-conditioned one. However, for a tightly tuned K_{opt}, feasible (and terminal) regions may not be large and the control law may not be robust.

7.3.5 Summary of dual mode algorithms and key points

One critically important issue not yet considered in this book is the impact of modelling accuracy. The focus so far has been on optimisation of nominal performance, but this is partially meaningless in the absence of an accurate model. Recall the observation that one cannot control better than one can model. Hence if the model is very inaccurate, what point is there in producing an *optimal* control law?

A highly tuned control law is not robust in general and hence the LQMPC algorithm may actually perform poorly on real applications if used naively. Industry largely uses a GPC/DMC (NTC) type of algorithm with small n_u but large n_y – this can reduce to low gain integral control but in the presence of significant plant uncertainty may give as good performance as one could reasonably expect, and with a relatively simple optimisation. In a later chapter we will consider robustness, but it should be noted that you are better to improve the robustness of wisely chosen control law that an over ambitiously tuned one.

Four algorithms have been given, SGPC, NTC, EUM and LQMPC. It is clear that LQMPC is the best algorithm on average but one might still think that none of them is entirely satisfactory. The question of which to choose, or modify, will however be process dependent. The designer will need to ask questions over the complexity of optimisation they can allow (for instance, how many constraint inequalities they can carry and how many d.o.f.), whether they want optimal performance or are happy with eliminating tracking offset using relatively low gain control. The tightness of constraints allied to limits on n_c places a restriction on the allowable terminal controller.

The conclusion we have reached, however, is somewhat surprising. In simple terms, MPC reduces to optimal control with constraint handling. The key advances since the 1960's where optimal control was not so readily accepted in industry are:

- The computing power is now available to do on-line optimisation of quite large quadratic programming problems.

- The evolution of MPC has produced a mechanism (dual mode MPC) for reformulating the infinite d.o.f. constrained optimal control problem as one with only a few d.o.f. Hence the optimisation is now tractable.

However, to get truly optimal control one may need a large number of d.o.f. and hence in many cases one may use a slightly suboptimal variant (for instance NTC) which could have a larger terminal region.

Summary: The choice of the mode 2 control law has a large impact on performance and feasibility. The choice must reflect model accuracy and control objectives.

7.4 Are dual mode-based algorithms used in industry?

Given all the recent academic developments in MPC, it is rather odd to note that industrial applications very rarely use state-space or transfer function models. Moreover, they very rarely use algorithms that can be related closely to optimal control and hence with a priori stability results. Why is this?

7.4.1 Efficacy of typical industrial algorithm

The most popular algorithm is essentially of the DMC form (Section 4.5) using the following guidelines:

- n_y is large and certainly greater than the system rise time.

- n_u is small, often just one.

- Models are usually step responses (e.g. DMC).

It should be emphasised (as argued in the previous two chapters) that these guidelines can ensure control that is close to linear quadratic optimal anyhow, at least for $n_u \geq 3$.

- If $n_y - n_u$ is greater than the system rise-time, then predictions will have all but settled and hence the difference on the numerical value of the cost between using an infinite and finite horizon is negligible. Given model uncertainty, such differences are irrelevant to practical control.

- For well-chosen sample rates, the principle dynamics of the input will have around 10 significant moves in a typical transient. In conjunction with the receding horizon implementation $n_u = 3$ is often enough to get close to this.

- For$^{\parallel}$ $n_u = 1$, DMC will give an open-loop form of response (at least if λ is small).

7.4.2   The potential role of dual mode algorithms

The basic MPC algorithm using finite horizons can give good performance, not with standing arguments on prediction mismatch given in Chapters 5 and 6. Hence one might wonder why bother with dual mode implementations. The simple answer is that there is no need for many cases, but:

1. The development of dual mode paradigms/infinite horizon algorithms solved a theoretical problem for the academics and hence gave MPC some analytic rigor. The insight gained is invaluable in understanding better the strengths and weaknesses of more typical MPC implementations.

2. When dealing with unstable open-loop processes or those with quite complex transient behaviour, a dual mode type of algorithm is likely to perform well with minimal tuning as the necessary insight is inbuilt.

> **Summary:** In practice the DMC/GPC algorithm is good enough to handle most industrial problems. Recent advances have given a better understanding of why this is so.

7.5   Advantages and disadvantages of the CLP over the open loop predictions

In many cases whether one uses an OLP or CLP implementation of dual mode control, it will make little practical difference apart from the issue of insight. This is

---

$^{\parallel}$We ignore here open-loop unstable plant and excessive nonminimum phase characteristics which need more precise tuning and hence higher n_u.

because if one minimises the same objective with the same d.o.f. (a reparameter-isation of the d.o.f. does not change the potential effects), then one must get the same answer. However, the reader will also be aware that the conditioning of different parameterisations can vary significantly. In general it is better to solve a well conditioned problem as opposed to a poorly conditioned problem.

Herein lies one major potential benefit of the CLP; it changes the conditioning of the optimisation problem. In cases where the OLP is poorly conditioned the CLP may not be and hence it will give more reliable answers. Of course one should add that the contrary could also occur and the reader may be best advised in general to check the conditioning with both approaches.

In this section we give a brief overview of some well understood problems which give useful insight to the control designer.

7.5.1 Cost function for optimal stable predictive control (LQMPC)

By optimal stable predictive control we mean the algorithm LQMPC of Section 7.3.4. In this section we will compare the cost function that arises from using the OLP and CLP. For simplicity of presentation the SISO case is used and comments on extensions to the MIMO case are given at the end.

7.5.1.1 LQMPC cost function with OLP

The cost, ignoring terms that do not depend upon the d.o.f., for an arbitrary dual mode algorithm was given in Section 6.6.2 as:

$$J = \Delta \mathbf{u}_{\rightarrow k-1}^T S \Delta \mathbf{u}_{\rightarrow k-1} + \Delta \mathbf{u}_{\rightarrow k-1}^T L \mathbf{x}_k \qquad (7.45)$$

where $S = H^T \mathrm{diag}(Q)H + \mathrm{diag}(R) + H_{n_c}^T P H_{n_c}$, $L = 2[H^T \mathrm{diag}Q P_x + H_{n_c}^T P P_{n_c}]$. What is clear is that there is a need to compute matrices H, H_{n_c}, P_x, P_{n_c} and then combine these to form matrices S, L. However, there is no useful insight into the conditioning of S, L and of the problem in general. The constrained optimum, that is, $\Delta \mathbf{u}_{\rightarrow k-1}^T = S^{-1} L \mathbf{x}_k$ may be easy to compute or not depending on the conditioning of matrix S.

7.5.1.2 LQMPC cost function with CLP

For the CLP the cost is given in Section 7.2.1 as

$$\min_{\underset{\rightarrow}{\mathbf{c}}} \; J = \underset{\rightarrow}{\mathbf{c}}^T S_c \underset{\rightarrow}{\mathbf{c}} + 2 \underset{\rightarrow}{\mathbf{c}}^T S_{cx} \mathbf{x} \qquad (7.46)$$

where S_c, S_{cx} have apparently complex definitions (7.20). However, it is known that for SISO LQMPC:

$$S_c = \mu I; \qquad S_{cx} = 0 \qquad (7.47)$$

with μ a scalar. One can prove this easily by noting a few things:

1. The unconstrained optimum $\mathbf{c}$ must be zero because:

    - By definition the control law minimising J in the absence of constraints is given as $\mathbf{u} = -K_{opt}\mathbf{x}$.
    - The LQMPC control law is given by $\mathbf{u} = -K_{opt}\mathbf{x} + \mathbf{c}$.

2. If the unconstrained optimal $\underset{\rightarrow}{\mathbf{c}} = 0$, then the cost function of (7.46) must be such that $S_{cx} = 0$.

    - The cost of (7.46) is quadratic.
    - If the optimum of a quadratic is the origin, then the cross terms must be zero.

3. The matrix $S_c = \text{diag}(\mu)I$ where μ is a scalar.

    - The cost function uses infinite horizons.
    - The impact on the cost of an identical perturbation now or at a later sampling instant must be identical.
    - One could compute μ but in the SISO case it is irrelevant.

Summary: For LQMPC, in the SISO case the objective function reduces to

$$J = \underset{\rightarrow}{\mathbf{c}}^T \underset{\rightarrow}{\mathbf{c}} \tag{7.48}$$

7.5.1.3 MIMO case and other terminal conditions

For arbitrary positive definite Q and MIMO systems a similar result will follow but in this case S_c will be block diagonal with each block the same.

One would compute the implied cost (6.11) for $\mathbf{c}_0 = 1$, $\mathbf{c}_i = 0, i \geq 1$ when $\mathbf{u} = -K_{opt}\mathbf{x} + \mathbf{c}$ and this would give $\hat{Q}$. For instance, given a state feedback, one can easily show that

$$\hat{Q} = B^T \Sigma B + R, \quad \Sigma - \Phi^T \Sigma \Phi = Q + K^T R K \tag{7.49}$$

Hence one can replace the optimisation of (7.46) with

$$J = \sum_{i=0}^{n_c-1} \mathbf{c}_{k+i|k}^T \hat{Q} \mathbf{c}_{k+i|k} \tag{7.50}$$

This is expected to be better conditioned optimisation than that given in (7.45), if nothing else because the level functions are spheres (in the SISO case) as opposed to ellipsoids or for the MIMO case ellipsoids of the input dimension N repeated n_c times as opposed to a $N \times n_c$ dimensional ellipsoid. Also the optimisation is centred at the origin.

Remark 7.8 *The same arguments may not cross over to EUM, NTC and SGPC implementations because there the objective is not solely to minimise the weighted norm of* $\underset{\rightarrow}{c}$*. That is, the underlying control law is not the optimum.*

Summary: Use of the CLP, especially with algorithm LQMPC, gives a well structured objective function with the unconstrained optimum on the origin.

7.5.2 Improved numerical conditioning with the CLP

The issue highlighted in this section appertains directly to an open-loop unstable plant and hence may be considered a rather special case. The reader is referred back to Section 5.4 for an illustration of the potential numerical difficulties caused by unstable processes. This is one case where the use of the CLP could be considered advisable.

7.5.2.1 Round-off errors

The difficulty with unstable processes is that the open-loop predictions diverge. Hence if one computes the prediction matrices, these rapidly become ill conditioned as the later rows contain very large values compared to the earlier rows. In the simplest case where only a few decimal places can be stored the important values containing the transients would be lost (even replaced by zero in the worst case) as the computer concentrated on storing the large values. Hence it is clear that MPC based on unstable predictions is unreliable for large output horizons simply because of round-off errors. Note, we have already discussed in Chapter 6 the issue of prediction mismatch which will inevitably occur in this case so this discussion is not repeated.

The advantage of the CLP is that it uses stable predictions in mode 1 and mode 2 and hence the prediction matrices do not contain coefficients with large variations in magnitude. This helps avoid issues with round-off errors as well as reduces prediction mismatch.

7.5.2.2 Conditioning errors

Let us suppose for a minute that the computer stores sufficient decimal places that round off in the prediction matrices is not an important issue. Reference to equation (7.45) shows that it is necessary to compute the matrices $S = H^T \text{diag}(Q)H + \text{diag}(R) + H_{n_c}^T P H_{n_c}$, $L = 2[H^T \text{diag}(Q)P_x + H_{n_c}^T P P_{n_c}]$. These involve multiplication of matrices that are already poorly conditioned. Moreover if matrix H has a range of coefficients over 10^n, then S is likely to have a range of coefficients over 10^{2n}, i.e. the problem is accentuated. Matrix S could therefore have very poor conditioning and yet the algorithm requires one further step equivalent to the inversion of S and then

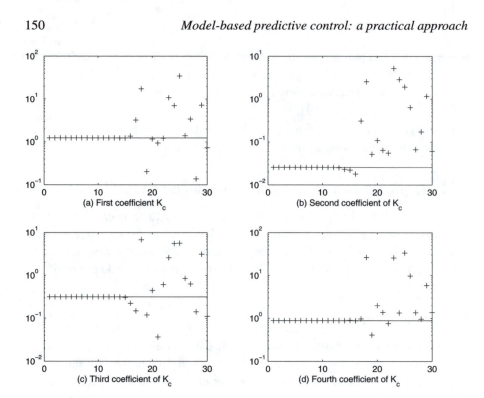

FIGURE 7.2

Dependence of norm of controller coefficients on n_c.

multiplication of L. The summary is that unless the computer stores several more decimal places than $2n$, the result could be seriously in error if not arbitrary.

These consequences are well illustrated in [104, 113] and similar plots are given here, in Figure 7.2, for convenience. The plots show the log of the absolute values of the coefficients of K_c of eqn.(7.23) which for LQMPC should be independent of n_c and therefore should be horizontal lines. The solid lines show the resulting coefficients using the CLP and the '+' denotes the coefficients using the OLP. It is clear that for small n_c where the computer has ample precision the two algorithms behave identically. However, as n_c increases the conditioning of the OLP approach becomes rapidly worse such that the control law cannot be computed reliably whereas CLP continues to behave well.

For information the example has state-space models

$$A = \begin{bmatrix} 2.6 & -0.05 & -0.5 & 1 \\ 1 & 0 & 0 & 0 \\ 0 & 1 & 0 & 0 \\ 0 & 0 & 0 & 1 \end{bmatrix}; \quad B = \begin{bmatrix} 1 \\ 0 \\ 0 \\ 1 \end{bmatrix}; \quad C = \begin{bmatrix} 1 & -2.2 & 1.12 & 0 \end{bmatrix}; \quad \lambda = 1$$

(7.51)

The optimal control law has coefficients

$$K = \begin{bmatrix} 1.2452 & 0.0256 & -0.3165 & 0.8827 \end{bmatrix}$$

(7.52)

Remark 7.9 *It should be noted that the poor performance does not arise until n_c is relatively large and hence one may therefore think that this result is irrelevant because practical guidelines suggest that n_c should be small. However, there are two counter arguments to this:*

1. *If approach 1 is known to be more poorly conditioned than approach 2, then approach 1 should not be used anyway and at best with caution.*

2. *LQMPC produces a difficult dilemma. If it is well tuned the terminal region maybe small and hence good feasibility necessitates the use of large n_c [64]. If one wants n_c small, then LQMPC may not be the best algorithm to choose.*

Summary: With unstable open-loop plant the reader is advised to use the CLP, especially if any horizons are large. Moreover, it is worth checking the conditioning of any implied computations.

7.5.3 Improved robust design with CLP

We will only give a brief overview of this issue here, as it forms the topic of a later chapter.

A major weakness of MPC is the need for an on-line optimisation (due to the constraints) at each sampling instant. As such there is no fixed linear control law and hence the approach does not appear to lend itself to traditional robust analysis and design**.

One potential benefit of the CLP is that the terminal control law can be hardwired (see Figure 7.1) and hence some design for robustness can be integrated into it. The perturbations used for constraint handling are then acting on a robustified loop and used wisely one can achieve a constrained control law with better robustness than will arise through the use of the OLP.

**Without going into computational demanding approaches such as those based on linear matrix inequalities and linear time varying representations, e.g. [59].

8

Constraint handling and feasibility issues in MPC

So far this book has dealt with the set up and tuning of MPC in the unconstrained case and this despite the fact that constraint handling is one of the major selling points of MPC. This is deliberate, as readers need to have a good feel for the underlying MPC algorithm before they are in a position to gain insight on the effect of constraints upon the optimisation.

A general observation is that the MPC algorithm optimises predicted behaviour subject to constraints being satisfied by those predictions. For instance, a racing driver optimises expected speed around a corner subject to staying on the track. However, this chapter introduces a more problematic element: what if the constraints cannot be satisfied? What do drivers do if the entry speed to the corner is so fast that they cannot stay on the track?

Summary: This chapter looks at how an MPC algorithm can be modified in the event that not all the desired constraints can be satisfied.

8.1 Introduction

The first questions the reader will need to know are *what is feasibility* and how is this term used in MPC?

8.1.1 Description of feasibility

Feasibility is usually a term applied to optimisation problems and describes whether a solution exists. For instance, consider the following quadratic program:

$$\min_{x} \; x^2 \;\; \text{s.t.} \;\; \begin{bmatrix} 1 \\ -1 \end{bmatrix} x \leq \begin{bmatrix} -1 \\ 0 \end{bmatrix} \tag{8.1}$$

This implies that both $x \leq -1$ and $x \geq 0$. Clearly these two are incompatible so it is not possible to find any values of x which satisfy the constraints; this problem is described as infeasible.

8.1.2   Feasibility in MPC

At times authors may use terminology such as algorithm X is infeasible, or approach Y is infeasible. Technically this is a misuse of language, but the authors simply mean that the optimisation problem that arises from the use of these algorithms is infeasible.

Alternatively, you may find the terminology that a given state **x** is infeasible. Once again this is a referred issue; i.e. the author is saying that if that value of state were to be used in the optimisation, then the optimisation would be infeasible. It is often convenient to use such shorthand terminology and the reader will become accustomed to it.

8.1.3   Overview of chapter

To some extent feasibility issues belong to the realm of the optimisation specialist and hence are not given a lot of space in this book. However, some of the concepts are particular to the MPC problem and these should be highlighted. Hence this chapter aims to answer questions such as:

- Why is feasibility important?

- What causes infeasibility in predictive control?

- How can one avoid or deal with infeasibility?

The reader should be aware that there are no simple answers to the problem in general and what is required is an awareness of the different compromises that are possible. The most suitable compromise can then be used for a given application. This chapter will discuss how infeasiblility arises and various mechanisms for avoiding it or tackling it.

Summary: Infeasibility implies that, for the current state, the constraints within the MPC algorithm cannot be satisfied. This chapter looks at mechanisms for overcoming or avoiding this.

8.2   Constraints in MPC

A discussion on feasibility requires clear assumptions on the constraints in the optimisation, as infeasibility is defined as an inability to satisfy all the constraints simultaneously.

8.2.1 Hard constraints

Hard constraints are constraints which must be satisfied. For instance, they may be limits on actuators or or on valves (which must lie between 0% and 100%) open. There is no point in a controller asking an input to go to a value beyond a hard constraint, as this cannot be achieved. What will occur is a mismatch between predicted behaviour and actual behaviour, the consequences of which could be arbitrary (see Chapter 6).

For this chapter define hard constraints as

$$C_H \underset{\rightarrow}{\mathbf{u}} + H_H \mathbf{x} - \mathbf{d}_H \leq 0 \tag{8.2}$$

where it is clear from the description that the constraints are state dependent as the predictions are state dependent.

8.2.2 Soft constraints

Soft constraints are those which should be satisfied if possible. For instance, there may be temperature or pressure limitations to prevent fatigue damage to equipment or to ensure quality. It is assumed that if necessary, soft constraints can be violated (ignored). Usually soft constraints are on outputs/states although they could also be applied to inputs. Such violations may have no effect on nominal stability results.

For this chapter define soft constraints as

$$C_S \underset{\rightarrow}{\mathbf{u}} + H_S \mathbf{x} - \mathbf{d}_S \leq 0 \tag{8.3}$$

8.2.3 Terminal constraints

These are somewhat artificial in that they arise from the control algorithm. The reader will recall that dual mode algorithms assume certain behaviour in mode 2. This assumption is only valid if the state during mode 2 is inside the associated maximal admissible set (MAS, see Section 11.9.1), this set being defined as the space within which soft and hard constraints are satisfied by the nominal control loop. The set into which the predictions are constrained is called a terminal set or a terminal region or terminal constraints or end point constraints.

Clearly terminal constraints are a form of soft constraint in that they are applied on far future predictions and arise from the desire for a guarantee of stability. However, they do in fact contain a mixture of hard constraints and soft constraints, this being obvious from the description of maximal admissible sets.

For this chapter define terminal constraints as

$$C_T \underset{\rightarrow}{\mathbf{u}} + H_T \mathbf{x} - \mathbf{d}_T \leq 0 \tag{8.4}$$

Summary: Constraints are a combination of hard, soft and terminal constraints. These are usually represented jointly in the simple form:

$$\underbrace{\begin{bmatrix} C_H \\ C_S \\ C_T \end{bmatrix}}_{C} \underset{\rightarrow}{\mathbf{u}} + \underbrace{\begin{bmatrix} H_H \\ H_S \\ H_T \end{bmatrix}}_{H} \mathbf{x} - \underbrace{\begin{bmatrix} \mathbf{d}_H \\ \mathbf{d}_S \\ \mathbf{d}_T \end{bmatrix}}_{\mathbf{d}} \leq 0 \qquad (8.5)$$

8.3 Why is feasibility important?

Before one leaps into a detailed discussion, the reader will be wondering why feasibility is such a big issue. This was partially answered in the example of Section 6.1.2 where it was shown that a mismatch between predictions and actual behaviour can lead to disaster.

8.3.1 Consequences of infeasibility

The basis of MPC is the optimisation of predicted performance subject to those predictions satisfying constraints. If the constraints cannot be satisfied, then optimisation is ill posed and therefore meaningless. In fact the optimisation itself is undefined and any answer will be arbitrary (some default). Arbitrary control decisions could cause random closed-loop behaviour and at worst instability.

Remark 8.1 *At times ([92, 100, 149]) saturation control* *is equivalent to the constrained optimal, but it is doubtful that this could be extended easily to a case with state constraints or the MIMO case with several interacting inputs and outputs. At the very least some comprehensive plant-dependent tests would be needed before saturation control was accepted.*

Summary: Without feasibility the MPC optimisation is ill posed and there is no assurance that the answer has any useful meaning.

8.3.2 Recursive feasibility

In general, an MPC optimisation being feasible now need not imply it will be feasible at the next sampling instant. There is a need in MPC for the algorithm to be feasible, at every sampling instant. This requirement is called recursive feasibility, that is:

---

*If the desired input exceeds its limit, simply set the input at the nearest limit.

> **Recursive feasibility:** *Feasibility now implies feasibility at the next sample and hence at all future samples.*

Within MPC, feasibility should really be defined as recursive feasibility but in practice the reader will have to read any document carefully to be clear on what is being implied.

By recursive feasibility, one is saying that one can set up the predictions so that any guarantees given now, can also be given at the next sampling instant (hence the word recursion). Failure to make such an assurance can allow infeasibility at some point in the future, at which point the MPC control law would become undefined. A simple illustration would be: driving your car with a strategy that keeps you on the road now, need not imply you stay on the road 50 yds later when you come to a sharp corner; hence your driving strategy was flawed. For a general assurance, your strategy must have infinite horizons and allow for all possible future scenarios.

8.3.2.1 The certain case

For the certain case recursive feasibility is relatively straightforward to establish, as it can be shown that the MAS (or any suitable invariant set) is a possible choice for the terminal region. One need only ensure that the tail (Section 6.3.1) is included in the prediction class and the arguments are then the same as those for guaranting stability where one falls back on the previous optimum predictions.

8.3.2.2 The uncertain case

The tail cannot be defined in the case with uncertainty and hence simple results do not apply. Consider an example where at each sample, a system is subject to a random but unknown bounded disturbance **b**. Set up an MPC strategy that ensures recursive feasibility.

- The one-step ahead prediction is given as

$$\mathbf{x}_{k+1} = \Phi\mathbf{x}_k + \mathbf{b}$$

- Assume the current nominal predictions are feasible if $\mathbf{x} \in S$ where S is as large as possible. However, due to $\mathbf{b}$, $\mathbf{x}_k \in S$ does not need to imply $\mathbf{x}_{k+1} \in S$ and hence one would not have recursive feasibility.

- Recursive feasibility is defined such that for some set S_f,

$$\mathbf{x}_k \in S_f \;\Rightarrow\; \mathbf{x}_{k+1} \in S_f. \; \forall \mathbf{b} \qquad (8.6)$$

and $\mathbf{x} \in S_f$ ensures constraint satisfaction.

- The set S_f may be far smaller than S.

- The description of such an S_f may be nontrivial in general [19, 40, 59] and moreover S_f may put severe limitations on allowable control moves.

Summary: An MPC algorithm is well-posed iff it has a guarantee of recursive feasibility. That is, any strategies proposed now cannot move you to a position where the algorithm becomes infeasible. This is easy to establish only for the certain case.

8.4   What causes infeasibility in predictive control?

Having accepted that infeasibility is undesirable and to be avoided, one needs to ask how infeasibility can arise. If one can identify the source, one is half way to producing an avoidance strategy. The following gives a brief overview of typical causes and possible mechanisms for avoiding them.

8.4.1   Incompatible constraints due to overambitious performance requirements

The most obvious cause of infeasibility is an incompatibility between soft and hard constraints. For instance if you want the internal house temperature to be 25 degrees when the external temperature is minus 25 degrees, it may simply be impossible given the maximum power output of the central heating. In this case you need to modify your objectives, that is, accept that the internal temperature will be cooler than desired.

Hard constraints dictate the most that can be delivered by a process. Soft constraints usually relate to what is desired. If the most desirable behaviour cannot be achieved with the components/actuators/etc. in place, then the desirable behaviour (soft constraints) must be modified to one that is achievable.

8.4.2   Conflicts with terminal mode control laws

Terminal constraints are artifical in that they are introduced to give a control law a guarantee of stability. However, they also ensure that the prediction class is feasible over an entire future horizon and so could be viewed as essential. If one cannot choose a prediction class that satisfies constraints over the entire horizon, then a problem is predicted to occur at some point.

If the terminal constraints are infeasible, then once again the optimisation is ill posed and hence the control trajectory that arises could be meaningless. One needs a strategy to ensure that the whole prediction class is feasible if one wants confidence in the control moves proposed by your MPC algorithm.

Fortunately this is not the whole story. A terminal constraint is based on an assumed terminal control law. If this is highly tuned, the associated terminal set may be small. However, there may exist an alternative terminal law with a far larger MAS and hence lack of membership of the given terminal region does not need to imply that a convergent and feasible sequence of control moves does not exist.

> **Summary:** If terminal constraints are infeasible, then replace with alternative terminal constraints associated to a terminal law with a larger MAS. To simply ignore them could give rise to an ill-posed optimisation.

8.4.3   Model uncertainty

Model uncertainty may cause infeasibility because the actual behaviour differs from the predicted behaviour. Hence, even though the nominal predictions could satisfy constraints over the entire future, a small change in the model will cause the actual behaviour to differ, and the associated predictions at the subsequent sampling instant could then violate constraints.

This issue is discussed briefly in Chapter 11 and is hard to deal with in general outside the use of a little back off. That is, do not drive the system so close to the limits that you have no flexibility left to deal with small variations from the expected behaviour.

One can form algorithms based on invariant sets (Section 11.11) to handle model uncertainty; however, the results are usually very conservative, as guarantees must allow for the worst case (which will arise with negligible probability). A more pragmatic approach is to accept that guarantees cannot be given where there is significant uncertainty and make other contingencies for the rare occasions where infeasibility arises.

> **Summary:** It is probably not good economics to use algorithms giving a guarantee of feasibility for the uncertain case, as the performance will be overly conservative.

8.4.4   Unstable open-loop processes

For the rare processes that are open-loop unstable one gets hard constraints on the outputs/states. That is, one can easily formulate scenarios whereby if an output goes outside a certain range, the system can no longer be stabilised. As such these systems must be treated with caution and it is paramount that the control strategy ensures recursive feasibility (eqn. (8.6)) of the most important constraints, including some allowance for uncertainty. This requires some rigorous analysis (e.g. [19]) beyond the scope of this book. The simplest way to view this is that the system should be constrained to lie within a certain space which still allows some slack in the control variables should it be needed.

8.5   Typical techniques for avoiding infeasibility

Having overviewed the typical causes of infeasibility one is in a position to propose strategies for dealing with them. We should however, make a key conceptual statement. If you want to be sure not to crash, either you do not drive at all or you drive very slowly – this may be an unacceptable solution.

In the following the assumption is taken that a guarantee of recursive feasibility is unrealistic in practice, although it may be reasonable for the nominal case. Hence, results are given for the nominal case and it is assumed that a sensibly tuned algorithm will deal with uncertainty.

The stronger a guarantee you want, the more conservative your control law will be. In practise there must be a compromise between feasibility assurances and performance.

8.5.1   Constraint softening

This is the most practically relevant strategy [93], but this book will not attempt to detail a precise algorithm, as there are too many process dependent decisions for this to be useful. Rather we will highlight the key issues and show how decisions can be included in an algorithm.

8.5.1.1   Simple strategy

If the constraint set (8.5) is inconsistent, then some constraints must be either relaxed or removed. A process dependent strategy could be developed using the following logic[†].

1. Relax (or remove) least important soft constraints and test for feasibility.

2. Relax (or remove) next most important constraints and test for feasibility.

3. etc.

The hope is that once enough soft constraints have been relaxed, the whole constraint set (8.5) will become feasible and one can continue. The decision making process is

---

[†]This relates to those constraints which are predicted to be violated. Relaxing nonactive constraints will change nothing.

taken by a supervisory controller before the constraints are downloaded to the MPC algorithm.

8.5.1.2 Relaxation based on infinity norms

The above strategy is a little crude in that the relaxation of a constraint is predetermined and hence may be more than required. An alternative is to define a variable $s \geq 0$ and replace the soft constraints by:

$$C_S \underline{\mathbf{u}} + H_S \mathbf{x} - \mathbf{d}_S \leq \mathbf{s} \tag{8.7}$$

The vector $\mathbf{s}$ defines the constraint violations (if positive). A possible strategy is then to minimise the maximum (weighted) value of $\mathbf{s}$ s.t. hard and terminal constraints. This will minimise the predicted soft constraint violations and is a relatively simple optimisation to set up (requiring only a linear program).

Remark 8.2 *The weakness of these approaches is that all the emphasis is placed on the soft constraints and none on the performance. A better strategy may wish to find some compromise between predicted performance and the violations of the soft constraints. The reader will also note that these issues are process dependent and outside the context of the MPC algorithm. Usually they would be taken by a supervisory controller.*

Summary: Feasibility would usually be ensured by a systematic relaxation of soft constraints. This would be determined at a supervisory level.

8.5.2 Back off and borders

This topic (e.g. [19, 40]) is mentioned in Section 8.3.2 above but is not a mainstream topic within MPC although it is used in the process industry. The basic idea is not to drive the system to the input limits, but to leave some extra freedom for emergencies. Typical examples may be when driving on the road, you never brake to the maximum or turn the steering wheel as far as you can; rather you leave a little extra to cope with the unexpected; i.e. the need to stop quicker or turn sharper. A racing driver takes the car right to the limit but as a consequence a small amount of uncertainty frequently results in the car leaving the track – they had no slack.

How much to back off the constraints and how exactly to incorporate this is process dependent. Clearly one should not do this unless necessary, as there will be a compromise with performance. Also one would expect no slack in the current control move (as you may need it all right now) and then increasing back off (slack) the further into the future of the predictions (where uncertainty is larger). The slack you allow yourself in the future predictions is used to deal with any differences between the predictions you make and the actual behaviour.

8.5.3   Simple illustration of back off

This illustration uses input limits only. Let the actual input limit be

$$\mathbf{u}_{k+i} \leq \bar{\mathbf{u}}, \quad \forall i \tag{8.8}$$

Now back off these constraints by a time dependent value $\mathbf{b}_i$; that is, let the MPC algorithm use constraints

$$\mathbf{u}_{k+i} \leq \bar{\mathbf{u}} - \mathbf{b}_i, \quad \forall i \tag{8.9}$$

It is assumed that $\mathbf{b}_i \geq \mathbf{b}_{i-1}$.

Theorem 8.1 *Using constraints (8.9) on the predictions ensures that extra control capacity is introduced at each sample to deal with uncertainty.*

Proof: At sampling instant k the constraints are given in (8.9). At sampling instant $k+1$ they will be given by

$$\mathbf{u}_{k+1+i} \leq \bar{\mathbf{u}} - \mathbf{b}_i, \quad \forall i \tag{8.10}$$

Comparing (8.9, 8.10) one sees that the back off on the constraint on prediction $\mathbf{u}_{k+i}$ has changed from $\mathbf{b}_i$ to $\mathbf{b}_{i-1}$; hence the predicted control $\mathbf{u}_{k+i}$ has an extra potential movement of $\mathbf{b}_i - \mathbf{b}_{i-1}$. $\square$

It is probably unwise to design the back off to give a guarantee of feasibility as the result will be too conservative. However some back off may improve the robustness of the feasibility assumptions in the presence of uncertainty. When producing invariant sets for uncertain systems (Chapter 11) the required back off can be automatically incorporated via the invariance conditions [19].

Summary: Using artificially tight constraints on future predictions automatically builds in some slack which can be used to retain feasibility in the presence of moderate uncertainty. The slack should be montonically increasing with the horizon.

8.6   Set point management and reference governor strategies

An obvious cause of infeasibility is due to a rapid set point change. This implies a large change in terminal constraint set (8.4) (due to a shift in steady-state values) and hence these may become inconsistent. The usual solution to this deployed in industrial software is to use an optimisation on top of the predictive controller. The role of this scheduling optimiser is to ensure that all set point trajectories are sensible. This is discussed no further, as the optimiser involves plant-wide knowledge and is application specific. Our interest is what to do when the optimiser fails to give feasible set points.

One popular algorithm for modifying set points to ensure they are feasible is reference governor [34] or management [36, 37, 38] strategies. These are motivated by 2 objectives: (i) computational simplicity and (ii) the integrity of an underlying controller. The idea is to modify the set point as required to ensure constraint satisfaction in the loop. Several ways of doing this exist and so it is best to read the literature if you want a comprhensive overview. This section simply summarises the main concepts and key insights.

8.6.1 Two reference governor algorithms

If feasibility is lost due to a set point change, then there is a fault in the requested strategy and asking for optimality is unrealistic. The main goal is to ensure a smooth transition back to feasibility with minimum perturbation to the nominal (and accepted) inner control loop.

With reference governor algorithms, the key philosophy is to send the MPC algorithm a set point w which is different from the true one r when changes in true set point would cause infeasibility. The controller set point w therefore has slower changes in value than the true set point. One might also argue that the simpler the strategy for regaining feasibility, the better. Two simple algorithms are given next.

Overview of algorithm: Let the set point at sample k be r_k and let the set point sent to MPC be w_k where it may be that $w_k \neq r_k$. Assume now that at the next sampling instant the set point changes to $r_{k+1} \neq r_k$. Select w_{k+1} to ensure feasibility of the inner loop control law. It remains then to ask how w is selected to ensure feasibility. Two simple possibilites are given next.

Exponential changes: Drive the MPC algorithm with a set point

$$w_{k+i|k} = w_k + [r_{k+1} - w_k](1 - \mu^{i-1}) \tag{8.11}$$

where $0 \leq \mu \leq 1$ and μ is chosen close enough to 1 to ensure feasibility. The closer μ is to one, the slower that w changes and therefore the slower the terminal set (8.4) changes.

Step changes only: Drive the inner loop with

$$w_{k+i|k} = w_k + [r_{k+1} - w_k]\beta; \quad 0 \leq \beta \leq 1 \tag{8.12}$$

This time the smaller β is, the slower w changes. β may be selected to optimise predicted performance or to maximise speed of convergence of w to r. If the desired set point is infeasible in steady state, then this algorithm will minimise the offset.

Remark 8.3   • *It may not be wise to drive w_{k+1} as close to r_{k+1} as possible, as this puts one on the limit of feasibility which has implications for robust stability. A little caution is always wise, as one wants to be sure of feasibility at the following sampling instant, given some uncertainty.*

- *Other reference governor algorithms are available in the literature for the interested reader. These are a bit more sophisticated and hence allow better optimality but at some increase in computation.*

Summary: Reference governor strategies can slow down requested set point changes to ensure the retention of feasibility and therefore controller integrity.

8.6.2   Links between the CLP and reference governor strategies

It should be noted that philosophically [119] there is actually little difference between reference management and MPC strategies and this becomes especially obvious when one uses the CLP. For instance consider [101, 103, 105] which are MPC strategies incorporating reference governor concepts.

- Both assume a fixed underlying loop controller. The design of this loop controller is not a central issue and is assumed given.

- Both manipulate the *set point* to this loop (see Figure 7.1) to ensure constraint satisfaction.

  1. Reference governor strategies use the set point variable itself.

  2. MPC based on the CLP uses perturbations to the set point.

- Both use knowledge of loop variables to *optimise* the *set point* and hence implicity introduce an outer loop.

- Both use predictions to check constraint satisfaction.

One might wonder then what the difference is, as these are usually presented as conceptually different. In fact the difference is simply in the parameterisation of the d.o.f. allowed for constraint handling.

Summary: Reference governor (management) strategies are equivalent to CLP MPC strategies. The differences are only in how the d.o.f. are parameterised.

8.7   Non-dual mode algorithms and feasibility

Often infeasibility is caused by the *artificial* terminal constaint (8.4) introduced to guarantee stability. This is an odd dichotomy; the very mechanism introduced to ensure stability could be the cause of instability (via infeasibility). Hence the price one pays for the guarantee of stability is a restricted region of applicability. That is, the algorithm is only defined in the space in which the terminal constraint is feasible.

Nondual mode based algorithms are those that do not deploy a terminal constraint beyond the usual assumption of the input going to a steady value after n_u steps. By avoiding the need for such a constraint, these algorithms can have [‡] a larger feasibility region; hence, rather perversely, they may in fact have better stability properties, despite the lack of a guarantee of stability.

One could argue that GPC with a large n_y and any n_u (except for open-loop unstable plant) will always give good and stable performance. If an accurate model is available, higher n_u will give better tuning but the difference with optimal control will probably be negligible for $n_u \geq 5$. Hence given that stability guarantees require terminal constraints and hence [64] large numbers of d.o.f., one may argue that practically one is just as wise to ignore dual mode algorithms and revert to GPC for practical industrial control where there is significant uncertainty.

A good discussion of options to ensure feasibility and guarantees of stability are in [64],[108]. The major underlying principle is to *define* a trajectory $\underrightarrow{u}$ which is known to be feasible w.r.t. input constraints (although it may not satisfy the terminal constraints) and which is guaranteed to give stable/convergent predictions. One can always use this trajectory when more optimal trajectories are infeasible. The definition of a feasible trajectory set is easy for stable plant (if suboptimality is accepted and there are no hard state constraints).

Summary: Terminal constraints (dual mode algorithms) give a stability guarantee but may also limit applicability to smaller regions than algorithms without terminal constraints. The terminal mode should be chosen wisely to avoid feasibility problems.

8.8 Summary

The MPC controller is only defined if the constraint set is feasible. Hence it is paramount that one sets up the strategy such that feasibility is always ensured.

One's wish list may be overambitious, for instance, the desire to achieve a optimal performance subject to tight limits on states. In this case there must be a systematic rule base that allows constraints or tracking requirements to be relaxed to ensure feasibility and therefore to ensure the consistency (well posedness) of the MPC law. The best means of relaxing requirements is process dependent and hence not discussed in detail in this book.

It is worthwhile identifying the main causes of infeasibility in a given process, for instance is it: (i) the terminal constraints; (ii) disturbances; (iii) over large setpoint

---

[‡]In particular this applies where there are input constraints only.

changes or (iv) soft constraints? This insight will direct the designer to the most appropriate formulation for the MPC algorithm.

Summary: The MPC algorithm is only well defined when constraints are feasible. Hence feasibility is essential.

9

Improving robustness – the constraint free case

It has been shown in the previous chapters that unconstrained* MPC gives a fixed control law. As such this law can be analysed to give loop sensitivities. The obvious question that arises is:

> Can MPC be tuned systematically to give a robust design?

The answer to this is both yes and no. This chapter will overview the common techniques available to improve the robustness of an MPC law and discuss how flexible they are. For authors interested in algorithms with a relatively low on-line computational burden there are two popular methods and most focus will be given to these.

The chapter is separated into sections introducing sensitivity functions and then descriptions and example implementations of the two popular approaches of: (i) the T-filter [158] and (ii) the Youla parameter approach [60].

Summary: This chapter demonstrates how the robustness of MPC can be improved in the absence of constraints.

9.1 Key concept used in robust design for MPC

For the reader interested only in understanding the key concepts, a brief summary is given next. Practitioners in MPC use what might be called a two d.o.f. or a two stage design. This draws on the work in the robust control literature, in particular, the Youla parameterisation [157]. Because of the particular structure of the MPC control law it is known that there is a decoupling between complementary sensitivity

---

*It is noted that constraint handling implies nonlinear control, hence sensitivity analysis is not strictly applicable and this is discussed later.

and other sensitivity functions, or in other words:

For the nominal case the robust design and design for performance are decoupled.

That is, one can tune the MPC law for *optimal performance* and then do a robust design as a separate stage without any effect at all on nominal performance [†], that is, the complementary sensitivity. There are many works looking at this, for instance [9, 21, 51, 60, 62, 66, 114, 150, 158] to name just a few.

9.2 Sensitivity functions for MPC with MFD models

In this section the sensitivity functions will derived for the MIMO case assuming an MPC controller structure and transfer function (MFD) models. In the SISO case there would be some simplification as commutativity is no longer an issue. These functions will be needed in later sections to analyse the impact on sensitivity of differing robust design methods.

9.2.1 Complementary sensitivity

Let the MPC control law and plant be given by

$$D\Delta u = Pr - Ny; \quad Ay = Bu \tag{9.1}$$

We will need the two alternative model and controller forms:

$$A^{-1}B = \tilde{B}\tilde{A}^{-1}; \quad D^{-1}N = \hat{N}\hat{D}^{-1} \tag{9.2}$$

Complementary sensitivity is the transference from the set point to the output. This can be computed by solving the model and controller equations of (9.1) as simultaneous equations. An example procedure for deriving the complementary sensitivity is given next.

$$\begin{aligned}
Ay &= BD^{-1}\Delta^{-1}[Pr - Ny] \\
(A\Delta + BD^{-1}N)y &= BD^{-1}Pr \\
(B^{-1}A\Delta + D^{-1}N)y &= D^{-1}Pr \\
(DB^{-1}A\Delta + N)y &= Pr \\
(D\tilde{A}\Delta\tilde{B}^{-1} + N)y &= Pr \\
(D\tilde{A}\Delta + N\tilde{B})\tilde{B}^{-1}y &= Pr \\
y &= \tilde{B}\underbrace{[D\tilde{A}\Delta + N\tilde{B}]^{-1}}_{P_c} Pr
\end{aligned} \tag{9.3}$$

_____

[†]However, it could be argued that the sensitivity achievable will depend on the original tuning.

Hence the complementary sensitivity is defined as

$$S_c = \tilde{B}P_c^{-1}P \tag{9.4}$$

where P_c defines the closed-loop poles:

$$P_c = [D\tilde{A}\Delta + N\tilde{B}] \tag{9.5}$$

If needed, the transference from set point to the input is given as $u = \tilde{A}P_c^{-1}Pr$.

9.2.2 Sensitivity functions used for robustness analysis

This section rapidly runs through the definitions of sensitivity to multiplicative uncertainty, to disturbances and to measurement noise.

9.2.2.1 Sensitivity to multiplicative uncertainty

In this case one is interested in robust stability, that is, for what range of uncertainty the closed-loop system is guaranteed to remain stable. Hence it is based on an analysis of the closed-loop poles. Closed-loop poles can be derived from

$$[I + GK] = 0; \qquad G = A^{-1}B, \qquad K = D^{-1}\Delta^{-1}N \tag{9.6}$$

Multiplicative uncertainty can be modelled as:

$$G \rightarrow (1+\mu)G \tag{9.7}$$

for μ a scalar (possibly frequency dependent). Substituting this into expression (9.6) the closed-loop poles for the uncertain case are derived from

$$0 = [1 + GK + \mu GK] = [1 + GK]^{-1}(1 + \mu GK[1 + GK]^{-1}) \tag{9.8}$$

As it is known that $[1 + GK]$ has stable roots by design (this is the nominal case), the system can be destabilised if and only if $|\mu GK[1+GK]^{-1}| \geq 1$. Hence the sensitivity to multiplicative uncertainty is defined as

$$\begin{aligned}
S_G &= [1 + GK]^{-1}GK \\
&= [G^{-1} + K]^{-1}K \\
&= [\tilde{A}\Delta\tilde{B}^{-1} + D^{-1}N]^{-1}D^{-1}N \\
&= \tilde{B}[D\tilde{A}\Delta + N\tilde{B}]^{-1}N \\
&= \tilde{B}P_c^{-1}N
\end{aligned} \tag{9.9}$$

Sensitivity to multiplicative uncertainty: MIMO case

$$S_G = \tilde{B}P_c^{-1}N \tag{9.10}$$

9.2.2.2 Disturbance and noise rejection

A disturbance (ζ) and noise (n) enter the system model as follows:

$$Ay = Bu + C\underbrace{\frac{\zeta}{\Delta} + An}_{f} \qquad (9.11)$$

For convenience these can be grouped as the term f. Redoing the algebra of eqn.(9.3) gives:

$$
\begin{aligned}
Ay &= BD^{-1}\Delta^{-1}[Pr - Ny] + f \\
(A\Delta + BD^{-1}N)y &= BD^{-1}Pr + f \\
(B^{-1}A\Delta + D^{-1}N)y &= D^{-1}Pr + B^{-1}f \\
(DB^{-1}A\Delta + N)y &= Pr + DB^{-1}f \\
(D\tilde{A}\Delta\tilde{B}^{-1} + N)y &= Pr + DB^{-1}f \\
(D\tilde{A}\Delta + N\tilde{B})\tilde{B}^{-1}y &= Pr + DB^{-1}f \\
y &= \tilde{B}P_c^{-1}[Pr + DB^{-1}f]
\end{aligned}
\qquad (9.12)
$$

Hence the sensitivities S_{yd}, S_{yn} of the output to disturbances/noise, respectively, are defined as

$$S_{yd} = \tilde{B}P_c^{-1}DB^{-1}\frac{C}{\Delta}; \quad S_{yn} = \tilde{B}P_c^{-1}DB^{-1}A \qquad (9.13)$$

Using a similar procedure the sensitivities S_{ud}, S_{un} of the input to disturbances/noise can be defined as:

$$S_{ud} = \tilde{A}P_c^{-1}NA^{-1}\frac{C}{\Delta}; \quad S_{un} = \tilde{A}P_c^{-1}NA^{-1}A \qquad (9.14)$$

Given the definitions above, it is easy to do a sensitivity *analysis*. The objective hereafter is to ask, how might one do a robust *design*?

Summary: The sensitivity functions for an MPC control law are straightforward to define.

Sensitivity to multiplicative uncertainty

$$S_G = \tilde{B}P_c^{-1}N \qquad (9.15)$$

Sensitivity to disturbances

$$\underbrace{S_{yd} = \tilde{B}P_c^{-1}D\Delta B^{-1}\frac{C}{\Delta}}_{\text{Output sensitivity}}; \qquad \underbrace{S_{ud} = \tilde{A}P_c^{-1}NA^{-1}\frac{C}{\Delta}}_{\text{Input sensitivity}} \qquad (9.16)$$

Sensitivity to noise

$$\underbrace{S_{yn} = \tilde{B}P_c^{-1}D\Delta B^{-1}A}_{\text{Output sensitivity}}; \qquad \underbrace{S_{un} = \tilde{A}P_c^{-1}N}_{\text{Input sensitivity}} \qquad (9.17)$$

9.3 T-filter approach

The use of the T-filter and how it alters the nominal control law was introduced in Sections 3.3.3, 4.3.4. This section looks in more detail on the effects the T-filter has on sensitivity. It should be emphasised that the major selling point of the T-filter is that one can modify sensitivity to parameter or signal uncertainty without any impact on the nominal tracking (complementary sensitivity); hence there is in effect a two stage design. First design tracking performance and then add a T-filter to improve sensitivity.

9.3.1 Overview

The use of a T-filter is the most popular approach to robust design in GPC. The prime reason is that the selection is based on intuition and hence is easily accessible to practising engineers; the T-filter can be considered as acting a little like a low-pass filter and hence is philosphically equivalent to a simple measurement filter taking out the high frequency noise.

Of course the downside of this is that the approach does not lend itself to systematic design [158]; that is, one cannot easily choose a T-filter to give a specified effect. Rather it is a *suck it and see* approach; i.e. choose a T-filter and try it. If the chosen T-filter does not achieve the desired sensitivity, it may not be obvious how to redesign it to improve matters. Some guidelines exist [158] but these are fairly simplistic and also do not cover all sensitivities, i.e. $S_G, S_{yd}, S_{yn}, S_{un}, S_{ud}$.

9.3.2 How a T-filter is included

$T(z)$ is treated like a design polynominal within the system model. Hence let the model be

$$a(z)y = b(z)u + T(z)\frac{\zeta}{\Delta} \qquad (9.18)$$

where T represents the colouring of the disturbance and noise signal (denoted f in equation 9.11). In practice the characteristics of f are not fully known; hence one could argue that T should be selected to best represent the key *disturbances* that one wants to reject using the well-known argument, '*the best way to reject disturbances is to include them in the system model*'. Of course the downside of this is that T will also affect the robustness to model parameter uncertainty and in a way that is not always beneficial. Hence it is not always best to use T to give the best disturbance model; that is, one may accept worse disturbance rejection to improve robustness to model uncertainty. Of course one is then brought back to the question, how can T be selected systematically?

Chapter 3 details the impact that a T-filter has on predictions (Section 3.3.3) and hence on the consequent control law (Section 4.3.4). In summary (for the SISO case) the control law of (9.1) is replaced by

$$\tilde{D}\frac{\Delta u}{T} = Pr - \tilde{N}\frac{y}{T} \tag{9.19}$$

Alternatively one could view that the individual compensators are replaced as follows:

$$D \to \frac{\tilde{D}}{T}; \quad N \to \frac{\tilde{N}}{T} \tag{9.20}$$

The impact on the sensitivity is transparent by direct substitution of (9.20) into eqns.(9.10, 9.13, 9.14) but noting that P_c is also redefined according to eqn.(9.5). For completeness the new sensitivity functions are derived next.
The equivalent of (9.5) is:

$$\tilde{P}_c = \frac{\tilde{D}\tilde{A}\Delta + \tilde{N}\tilde{B}}{T} = P_c \tag{9.21}$$

Hence substituting (9.20, 9.21) into (9.10, 9.13, 9.14) gives:

$$S_G = \tilde{B}P_c^{-1}\frac{\tilde{N}}{T} \tag{9.22}$$

$$S_{yd} = \frac{\tilde{B}P_c^{-1}\tilde{D}\Delta B^{-1}C}{\Delta T}; \quad S_{yn} = \frac{\tilde{B}P_c^{-1}\tilde{D}\Delta B^{-1}A}{T} \tag{9.23}$$

$$S_{ud} = \frac{\tilde{A}P_c^{-1}\tilde{N}A^{-1}C}{\Delta T}; \quad S_{un} = \frac{\tilde{A}P_c^{-1}\tilde{N}}{T} \tag{9.24}$$

Because the controller parameters change (see eqn.(9.20) and Section 4.3.4) one cannot easily compare the new and old sensitivities except by direct computation; that is, using an a posteriori analysis. However, it is clear that T appears in the denominator of many of the new sensitivities and hence there is an expectation that if $1/T$ is a low pass filter, then sensitivity will be improved at high frequencies. This expectation is largely realised in practice.

Remark 9.1 *A typical choice of T is* $(1 - 0.8z^{-1})^n$, $n = 1$ *or* 2. *This choice is intuitive in that if sampling at about 1/10 of the rise time[‡], then a typical dominant process pole would be around 0.8. Hence this is a sensible pole for a low-pass filter on output measurements.*

Summary: The use of a T-filter is simple and moreover recognised as essential in many real applications to reduce the input sensitivity to high frequency noise. It often has little effect on the output sensitivity but has a noticeable effect on input sensitivity S_{un}.

[‡]This is a common guideline for MPC.

9.3.3 Illustration of the T-filter

This section will be used to illustrate the impact of the T-filter on sensitivity for some simple examples.

9.3.3.1 Example 1

Consider the following example and design an MPC control law with two possible T-filters

$$G(z) = \frac{z^{-1}}{(1 - 0.9z^{-1})(1 - 0.5z^{-1})}; \quad \begin{cases} T_1(z) = 1 - 0.8z^{-1} \\ T_2(z) = (1 - 0.8z^{-1})^2 \end{cases} \quad (9.25)$$

The sensitivity functions are plotted in Figure 9.1 where the solid line represents $T = 1$, the dashed line $T = T_1$ and the dotted line $T = T_2$. The inclusion of the T-filter has given good reductions in every sensitivity function over the high frequency range and this bears a strong link to the amount of implied filtering.

- S_G is actually better over the whole frequency range so T has improved robust stability.

- Output sensitivity is *worse* at mid and low frequencies and better at high frequencies. One might argue that at low frequencies the sensitivity tends to zero anyway (due to the integral).

- The input sensitivity is better over the whole frequency range.

- The reader is reminded that the above changes come at no cost to the nominal tracking performance.

9.3.3.2 Example 2

Consider the example below and two possible T-filters

$$G(z) = \frac{z^{-1} - 1.3z^{-2}}{(1 - z^{-1} + 0.8z^{-2})}; \quad \begin{cases} T_1(z) = 1 - 0.8z^{-1} \\ T_2(z) = (1 - 0.8z^{-1})^2 \end{cases} \quad (9.26)$$

The sensitivity functions are plotted in Figure 9.2 where the solid line represents $T = 1$, the dashed line $T = T_1$ and the dotted line $T = T_2$.
In this case the T-filter seems to have made the sensitivity far poorer especially with strong filtering.

- S_G is worse over what is most likely to be the critical frequency range and better only at very high frequencies.

- Output sensitivity varies between better and worse with and without filtering and there is no conclusive statement that can be made.

- The input sensitivity is better over the high frequencies and worse over mid and low frequencies.

9.3.3.3 Examples with the MIMO case

Due to the potential interactive effects it is not so easy to illustrate the impact of a T-filter on a MIMO process. However, one could expect similar trends, that is a reduction in input sensitivity (S_{un}) to noise possibly at the expense of poorer sensitivity elsewhere.

More importantly one should note that if a systematic design of T is difficult for the SISO case, it is far harder for the MIMO case where $T(z)$ could become an MFD. To the authors knowledge no systematic rule base exists beyond the obvious choice of some 'logical' low pass filtering. As a consequence no illustrations are given here.

Summary:

1. Although intuitive arguments suggest that the T-filter will improve sensitivity, this is not always the case. It is paramount that the user do some analysis before implementing on a real process.

2. In many cases it will be observed that a T-filter trades off sensitivity in different frequency bands; that is, pushing sensitivity down at high frequency can push it up at low or intermediate frequencies.

9.4 Youla parameter approaches

This section introduces a more systematic method for improving robustness. The downside of this is that the design is no longer intuitive and requires both optimisation and more precise specification of objectives.

9.4.1 Introducing a Youla parameterisation into a MPC control law

The Youla parameterisation [157] has been used as a tool to improve the robustness of fixed linear control loops and is well known in the robust control literature. In essence the minimal order control law is modified with a d.o.f. (denoted here as the Youla parameter Q). If incorporated in a particular way, this Q can be shown to alter the system sensitivity functions without the altering nominal performance (complementary sensitivity). In fact the T-filter approach of the previous section is in effect a form of Youla parameterisation, as the complementary sensitivity is unaffected by the change of controller due to T. In this section it will be shown how the effect of the Youla parameter upon sensitivity can be made more transparent and hence d.o.f. can be made available for a systematic robust design.

For an MPC control law (9.1) the Youla parameter Q can be introduced by modifying the compensators as follows:

Youla parameterisation (MIMO case):

$$D \rightarrow D - QB$$
$$N \rightarrow N + QA\Delta$$

(9.27)

Substitution of (9.27) into (9.3) will modify the complementary sensitivity as follows:

$$y = \tilde{B}[D\tilde{A}\Delta + N\tilde{B}]^{-1}Pr$$

$$y \rightarrow \tilde{B}\underbrace{\left[(D - QB)\tilde{A}\Delta + (N + QA\Delta)\tilde{B}\right]^{-1}}_{P_{cq}} Pr$$

(9.28)

However it can be shown that $P_{cq} = P_c$ and therefore complementary sensitivity are actually independent of Q. For instance:

$$\begin{aligned}
P_{cq} &= (D - QB)\tilde{A}\Delta + (N + \Delta QA)\tilde{B} \\
&= D\tilde{A}\Delta + N\tilde{B} - QB\tilde{A}\Delta + \Delta QB\tilde{B} \\
&= D\tilde{A}\Delta + N\tilde{B} \\
&= P_c
\end{aligned}$$

(9.29)

Summary: Nominal tracking/performance is unaffected by Q, that is, by the controller modification of (9.27).

9.4.2 Effects of the Youla parameter on sensitivity

It has been noted that one can modify D, N without affecting nominal tracking. However, Q has a marked effect on the sensitivity functions and this is discussed next. With little further explanation we will simply substitute the control parameterisation (9.27) into the sensitivities given in (9.10, 9.13, 9.14). What is most notable is that the sensitivity functions are affine (linear) in Q. Hence there is a direct correspondence between Q and sensitivity which can be used in a systematic design procedure [60].

Sensitivity to multiplicative uncertainty:

$$S_G = \tilde{B}P_c^{-1}N \quad \rightarrow \quad \tilde{B}P_c^{-1}[N + QA\Delta] \qquad (9.30)$$

Output sensitivity:

$$\begin{aligned} S_{yd} &= \tilde{B}P_c^{-1}D\Delta B^{-1}\tfrac{C}{\Delta} \quad \rightarrow \quad \tilde{B}P_c^{-1}(D - QB)\Delta B^{-1}\tfrac{C}{\Delta}; \\ S_{yn} &= \tilde{B}P_c^{-1}DB^{-1}A \quad \rightarrow \quad \tilde{B}P_c^{-1}(D - QB)B^{-1}A \end{aligned} \qquad (9.31)$$

Output sensitivity:

$$\begin{aligned} S_{ud} &= \tilde{A}P_c^{-1}NA^{-1}\tfrac{C}{\Delta} \quad \rightarrow \quad \tilde{A}P_c^{-1}(N + QA\Delta)A^{-1}\tfrac{C}{\Delta}; \\ S_{un} &= \tilde{A}P_c^{-1}NA^{-1}A \quad \rightarrow \quad \tilde{A}P_c^{-1}(N + QA\Delta) \end{aligned} \qquad (9.32)$$

As the sensitivities are affine in Q, optimisation over Q to minimise sensitivity is straightforward. These ideas were introduced into MPC in [60] and many other authors (e.g. [9], [51]) found subsequent applications. Here we re-emphasise the key points are that predictive controllers are such that:

Summary:

1. The controller denominator is wholly in the forward path and the controller is wholly in the return path.

2. This separation causes the sensitivity functions to be linear (affine) in Q.

3. Q has no effect on nominal tracking (complementary sensitivity).

9.4.3   Optimising sensitivity

The dependence on Q is affine and therefore it is easy to manipulate/select Q to have the most desired affect. However, herein lies a caveat; one needs a precise numerical objective. In the robust control literature it is normal to define weighting functions which implicitly set this objective. Such a procedure can be followed and a possible mechanism is detailed in [51]. In line with the philosophy of the rest of this book, however, we will illustrate how more simple minded algorithms can give effective results. The following algorithm could be coded and understood in just a few minutes.

Algorithm 9.1   *1. Let the given sensitivity function be*

$$S_i(z) = G_i(z) + H_i(z)Q(z) \qquad (9.33)$$

    2. Define

$$Q(z) = Q_0 + Q_1 z^{-1} + \cdots + Q_n z^{-n} \qquad (9.34)$$

3. *Compute the frequency response of $S_i(z)$ at a set of frequencies ω_i and stack in a vector* **s** *as*

$$
\mathbf{s} = \begin{bmatrix} S(\omega_1) \\ S(\omega_2) \\ \vdots \\ S(\omega_n) \end{bmatrix} = \begin{bmatrix} G(\omega_1) \\ G(\omega_2) \\ \vdots \\ G(\omega_n) \end{bmatrix} - F\mathbf{Q}; \quad \mathbf{Q} = \begin{bmatrix} Q_0 \\ Q_1 \\ \vdots \\ Q_n \end{bmatrix} \tag{9.35}
$$

where the details of how to compute F are omitted as cumbersome to write although trivial in structure[§].

4. *Select* **Q** *as follows[¶]:*

$$
\min_{\mathbf{Q}} \quad \|W\mathbf{s}\|_r \tag{9.36}
$$

where typically $r = 1, 2$ or ∞ and W is a weighting matrix chosen to emphasise given frequency ranges.

The next section will illustrate this algorithm and its efficacy.

Remark 9.2 *If there is a desire to minimise several sensitivity functions, the appropriate vectors can be stacked within* **s**. *This does not increase the complexity of the algorithm beyond the increase in the number of rows in* **s**.

Summary: One can use a simple optimisation to minimise the sensitivity w.r.t. the parameters of $Q(z)$ as a suboptimal alternative to a rigorous robust design.

9.4.4   Use of Q on examples 1 and 2

Take example 2 from Section 9.3.3.2 and design a 10-term Q to minimise the 2-norm of **s** contructed from S_{yd} and S_{ud} (with unity weighting, i.e. $W = 1$). The corresponding sensitivities (in dash-dot lines) are given in Figure 9.3 and are overlaid with the T-filter results of Figure 9.2.

The results do not look much different from those obtained by the T-filter (dashed and dotted lines) in this case. However, some important points should be made:

1. The result is optimum w.r.t. the objective given (which was simplistic).

2. One has the option of giving a more specific objective:

   • Use frequency weighting to emphasise certain frequencies.

[§] One wants the *i*th row of $F\mathbf{Q} = H(\omega_i)Q(\omega_i)$. So $F_{i,k+1}$ contains terms of the form $e^{jk\omega_i T}$, T the sampling period.

[¶] To ensure real answers one should include matching positive and negative frequencies in **s**.

- Use an infinity norm to minimise the worst case.

3. Neither of these are true with the T-filter with which the consequent sensitivity is hard to control.

9.4.5 Example 3

This example [114] is a good illustratation of the potential benefit of the Youla parameterisation. It comprises a set of three plants of which the first is the nominal model from which the basic MPC controller will be computed. The denominator will be taken to be constant:

$$A(z) = 1 - 1.8z^{-1} + .27z^{-2} + .27z^{-3}$$
$$B(z) = z^{-1} + 0.1z^{-2} - 0.2z^{-3} \tag{9.37}$$

A set of possible numerator perturbations are given as:

$$\delta b_1 = z^{-1}[0.533z^{-1} - 0.1333z^{-2} - 0.333z^{-3}]$$
$$\delta b_2 = z^{-1}[-0.35z^{-1} - 0.035z^{-2} + 0.07z^{-3}] \tag{9.38}$$
$$\delta b_3 = z^{-1}[0.3404z^{-1} + 0.0851z^{-2} + 0.2128z^{-3}]$$

so that $B(z) = b_0(z) + \lambda \delta b_1 + \lambda_2 \delta b_2 + \lambda_3 \delta b_2$, $|\lambda_i| \leq 1$, $\sum |\lambda_i| \leq 1$. The nominal controller is not robustly stable over the whole class of possible numerators. However, after addition of a Youla parameter stabilisation is achieved. Figure 9.4 shows plots of the sensitivity function S_G of eqn.(9.30) against frequency $\omega T = 0, ..., 2\pi$ for a 3rd, 10th and 20th order $Q(z)$. It is clear that Q has allowed a significant, and systematic, improvement in the sensitivity, which becomes close to flat (except at low frequencies due to the integrator) for large orders as expected.

Summary: The Youla parameter approach allows a systematic minimisation of sensitivity and hence is a useful tool for ensuring robust stability.

9.4.6 Potential improvements and comments

- In general one may wish Q to be a transfer function. Unfortunately the simplest computations, as given here, imply that Q is restricted to a FIR; denominator terms imply a more complex optimisation or more algebra. It is possible of course to find a high order FIR and then find a transfer function approximation, or to fix the denominator a priori using some form of low pass guidelines as in T-filter design.

- A systematic approach, based on the robust literature, for selecting Q is given in [52] and [9, 10]. However, the selection of weighting or objective functions as required for a rigorous H_∞ algorithm is often a little arbitrary and could nullify the supposed benefits; it requires good experience with the algorithms so is not for the novice. Approaches limited to FIRs are available in [60, 62].

- There is an integrator in the control law and hence a difference operator Δ in S_G (eqn. 9.15). This means that many of the standard results in the robust literature will need modification (for instance, one might use $\Delta = 1 - (1 - \varepsilon)z^{-1}$), as they do not apply when there are poles/zeros on the stability boundary.

- It is well known that the optimum Q to minimise the ∞-norm of S_G gives a flat spectrum (where there is no Δ) and hence *an analytic result* can be given for Q where there is no integral.

This author favours the simpler FIR based approach where this gives good results, as this is straightforward to code and understand. However one obvious weakness of a restriction to FIR models comes across clearly in the Figure 9.3 where one can see spiky sensitivity. A good compromise solution would be to specify a given set of poles for $Q(z)$; for instance, as one would via the T-filter approach, and then add the Youla parameter after this. A two stage Youla parameterisation could be given as: (i) first select an *optimum* T-filter and find the corresponding control law and then (ii) select an optimum Q to improve sensitivity in targeted frequency bands, i.e.

$$
\begin{array}{ccccc}
\text{Initialise} & & \text{Include T – filter} & & \text{Add Youla} \\
\hline
D & \rightarrow & \dfrac{\tilde{D}}{T} & \rightarrow & \dfrac{\tilde{D} - BQ}{T} \\
N & \rightarrow & \dfrac{\tilde{N}}{T} & \rightarrow & \dfrac{\tilde{N} + A\Delta Q}{T}
\end{array}
\tag{9.39}
$$

Hence the T-filter is used to give good high frequency noise rejection via the low-pass characteristics and then Q is used to meet more general sensitivity objectives as required over the mid frequency range.

> **Summary:** An effective and simple strategy would take the T-filter based controller as a start point and use the Youla parameterisation approach to further improve sensitivity if required.

9.5 Internal model approaches

This section is deliberately brief and indicates methods which may be more popular with users of state-space models, as in this case the form of the Youla parameterisation would be different. Moreover the Youla parameterisation is likely to be incorporated into the state observer. One could argue the case for a Kalman filtering approach, as this is supposedly optimal, in conjunction with the nominal control law.

However, there are two weaknesses to this argument. First this assumes knowledge of the signal characteristics which usually have to be estimated and secondly it does not cater for all the sensitivity functions (such as S_G) but rather gives optimal state estimation for a given signal uncertainty. The reader is referred to other more comprehensive literature such as that on robust control for a fuller description of these approaches as they need not be restricted to an MPC context.

A simpleminded (but not rigorous) approach which could be linked to an observer based control design is based on the internal model principle (e.g. [30]) and is outlined next.

9.5.1   Nominal case

Let the control law be represented as

$$u = Ky + Pr \tag{9.40}$$

(The observer is implicit in K, P.) Now assume there is a model G_m of the real process G. Simulate G_m alongside the real process and compute the error between the two so that

$$e = Gu - G_m u = y - G_m u \tag{9.41}$$

Use this error to generate a perturbation δu to the control law.

$$\delta u = -Qe = -Q(y - G_m u) \tag{9.42}$$

The overall closed-loop is now represented by two equations

$$y = Gu; \quad u = Ky + Pr + \delta u = Ky + Pr - Q(y - G_m u) \tag{9.43}$$

Clearly if $G = G_m$, Q does not appear in the nominal closed-loop dynamic, hence Q introduced as in (9.42) is a form of Youla parameterisation.

9.5.2   Uncertain case

Now let the true process by uncertain. Using a multiplicative uncertainty description gives $G = G_m(1 + \mu)$. Substitute this into (9.43) and rederive the closed-loop equations as follows:

$$\begin{aligned}
(I + Q(G_m[1 + \mu] - G_m) - KG_m[1 + \mu])u &= Pr \\
(I + \mu Q G_m - KG_m[1 + \mu])u &= Pr \\
(I + \mu Q G_m - KG_m[1 + \mu])u &= Pr \\
(1 - KG_m + \mu[QG_m - KG_m])u &= Pr
\end{aligned} \tag{9.44}$$

The sensitivity function S_G is therefore given as

$$S_G = (1 - KG_m)^{-1}[QG_m - KG_m] \tag{9.45}$$

which is affine in Q. Once again the Youla parameter can be used to minimise sensitivity in a straightforward manner. However, one cannot simply choose $Q = K$, as this would remove all output feedback from the control law.

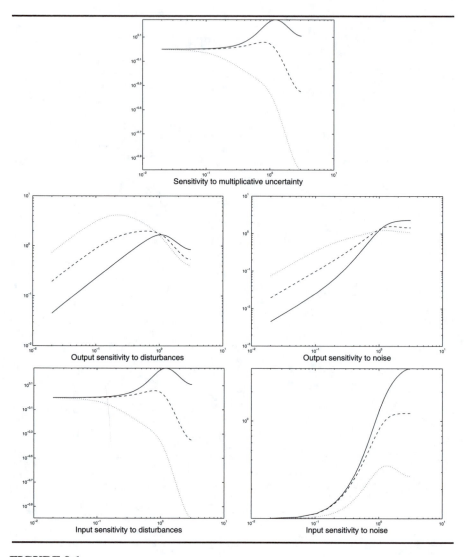

FIGURE 9.1
Sensitivity for example 1.

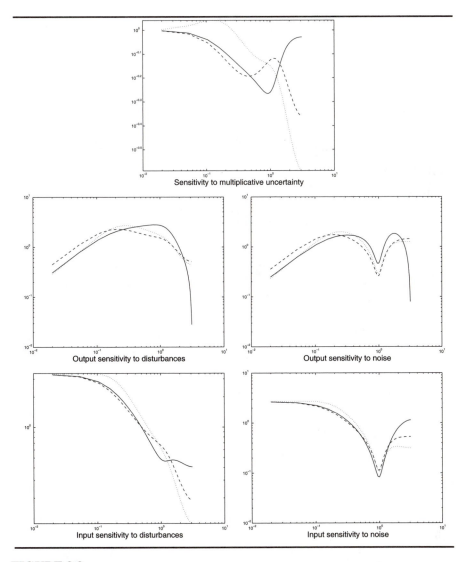

FIGURE 9.2

Sensitivity for example 2.

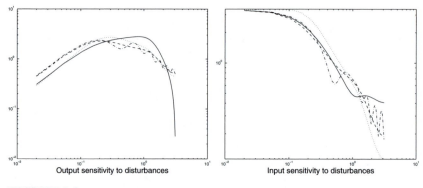

FIGURE 9.3

Sensitivity for example 2.

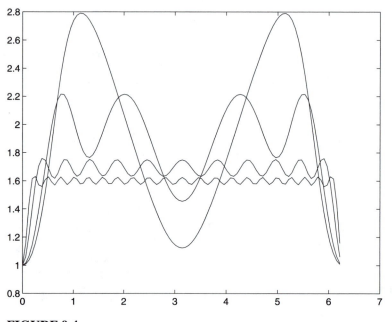

FIGURE 9.4

Sensitivity functions for $n_Q = 0,\ 3,\ 10,\ 20$.

10

The relationship between model structure and the robustness of MPC

The purpose of this chapter is to highlight the importance of the modelling stage and assumptions. There are two main aspects to this:

1. Models are used to form system predictions and prediction accuracy is fundamental to the efficacy of the control design.

2. Models implicitly contain a disturbance model. The accuracy of this model has an impact on the accuracy of prediction and therefore the consequent behaviour.

This chapter will study the two aspects separately, that is, how the modelling stage affects the prediction errors and also how assumptions on the model structure affects the implied sensitivities corresponding to the resulting control law.

Summary:

- It is useful to be aware of the impact on sensitivity of modelling assumptions.

- Although strictly speaking this chapter deals only with the constraint free case, because the insights are based on predictions, one might expect similar insights to apply to the constrained case.

10.1 Introduction

It will have been noted in the first few chapters that MPC can make use of many model structures of which the most popular are FIR and transfer function or state space. Within the choice of transfer function models there are numerous choices due to the different noise models that can be adopted although these are usually summarised by some design freedom such as in the T-filter. The question that is of interest to the control designer is to understand the pros and cons associated to each model choice.

This chapter will look in detail at the impact the model structure has on the control law and hence on sensitivity. The reader is reminded that historically industrial variants of MPC favoured FIR models (which were easy to understand but had large numbers of coefficients) whereas academics favoured transfer function/state-space models (which were more compact and easier to analyse). One might be wondering whether there is some reason for the industrial preference, over and above the ease of interpretation of step responses, or perhaps whether there are actually disadvantages to the use of FIR models.

Unsurprisingly this chapter has a sizable overlap with the previous chapter due to the links with observer (e.g. the T-filter) design. The aim is to analyse this and hence to gain some insight. The key thing that will become apparent and was clear from Chapter 3 is that different models imply the use of different predictors which hence give different prediction errors. This in turn has an impact on performance.

> **Summary:** The use of different prediction models implies the use of different observers (or T-filters/disturbance models) and hence gives rise to control laws with different loop sensitivity.

10.1.1   Importance of prediction errors

For convenience this chapter uses the *hat* $(\hat{\cdot})$ notation to denote both a current measurement and a prediction.

In predictive control design the optimisation of the chosen cost depends explicitly on the system predictions $\hat{y}_{k+j|k}$. Therefore it is not unreasonable to expect that the quality of control is closely correlated to the accuracy of prediction.

In reality one is minimising terms of the form $\hat{y}^2_{k+j|k}$ when in fact one wants to minimise the true and unknown output prediction $y_{k+j|k}$. Let e denote the prediction error, then the actual output is given as $y_{k+j|k} = \hat{y}_{k+j|k} + e_{k|j+k}$. Hence the error in the square of the output prediction is:

$$y^2_{k+j|k} = [\hat{y}_{k+j|k} + e_{k|j+k}]^2 = \hat{y}^2_{k+j|k} + \underbrace{e^2_{k|j+k} + 2[\hat{y}_{k+j|k}][e_{k|j+k}]}_{Error\ in\ J} \tag{10.1}$$

On average the sign of the error term $[\hat{y}_{k+j|k}][e_{k|j+k}]$ will be zero mean. Hence to give a synergy [126] between the modelling and the control design (that is, to minimise errors in the performance index), one models with the aim of forming the predictions which minimise $e^2_{k|j+k}$. That is, the model is chosen to minimise the effect of measurement noise and unmeasurable disturbances in introducing prediction errors into the cost J. One would expect that on average this modelling strategy would minimise the effect of such errors on the resulting control law*.

---

*It should be noted, however, that some researchers are looking more closely at the impact of minimising the cross term.

10.1.2 Overview of chapter

The chapter is arranged as follows. First the predictions associated to common models are analysed to make inferences about the variance of the prediction errors. Then some sections illustrate through examples the prediction errors and sensitivity functions that arise with the use of different models. Parameter uncertainty is excluded from the discussions in this chapter.

10.2 Summary of models to be compared

Let $G(z)$ denote the system model such that a generic process can be represented as:

$$y(z) = G(z)u(z) + F(z)d(z); \quad \hat{y}(z) = y(z) + \zeta(z) \tag{10.2}$$

where $\hat{y}$ denotes the measurement, the term $d(z)$ represents disturbance effects and $\zeta(z)$ represents noise effects. We will use the case where $d(z)$ is integrated and filtered white noise; that is, it has the form $d(z) = v(z)/\Delta(z)$, $\Delta(z) = (1 - z^{-1})$, where v, ζ are both unknown and zero mean. This model allows for nonzero steady state disturbances on the system state.

For the purposes of this chapter it is assumed that $F(z) = 1$; clearly the conclusions will reflect this assumption.

10.2.1 Prediction models and prediction errors

For clarity of presentation, this section gives a brief summary of the prediction models derived in Chapter 3. We will consider the nominal case (no parameter uncertainty) only, so that in the case where $v = 0, \zeta = 0$, exact predictions are possible. $\hat{y}$ denotes both measured output and predictions based on measured data.

Summary: The aim of this section is to look at the prediction errors inherent in prediction models due to measurement noise.

10.2.1.1 Prediction and prediction errors with FIR models

The development here is a brief summary of that available in Section 3.5.2. The step response $H(z) = G(z)/\Delta(z)$, $\Delta(z) = 1 - z^{-1}$ gives a prediction model:

$$y(z) = H(z)\Delta u(z) + \frac{y_0}{1 - z^{-1}} + \frac{v_f(z)}{\Delta(z)} \tag{10.3}$$

where y_0 is an unknown initialisation point to cater for the operating point and past disturbances and $v_f(z)$ corresponds *future* (unknown) disturbances only. This model is equivalent to (10.2) if $F(z) = 1$ and otherwise introduces bias into the predictions.

Given that the term v_f is unknown (and zero mean) one can write the difference equation form of (10.3) at two different sampling instants and eliminate y_0. Hence the prediction is given from:

$$y_{k+i|k} = \hat{y}_k + \sum_{j=0}^{i-1} H_{i-j}\Delta u_{k+j|k} + \sum_{j=1}^{n} [H_{j+i} - H_j]\Delta u_{k-j} \qquad (10.4)$$

The true prediction requires $n = \infty$; however, $\lim_{j\to\infty}[H_{j+i} - H_j] = 0$, hence for large enough n, the truncation errors will be minimal. In this section truncation errors are ignored.

- The main cause of error therefore is through the measurement $\hat{y}_k$ which is corrupted by noise ζ_k (i.e. $\hat{y}_k = y_k + \zeta_k$).

- Past disturbances are included exactly through y_0 when $F(z) = 1$.

- Therefore prediction errors are dominated by $\zeta(z)$, unscaled and unfiltered and future unknown v_k. Assuming that on average $v(z) = 0$, the variance of such prediction errors is therefore equal to the variance of ζ_k.

Summary: Assuming no truncation errors, one can approximate the variance of the prediction errors for step response models as

$$\text{var}(e_{k+i|k}) \le \text{var}(\zeta_k) = \sigma^2 \qquad (10.5)$$

10.2.1.2 Prediction and prediction errors with transfer function models

This section gives a brief summary of results available in Section 3.3.3.3. A typical transfer function model used for prediction is

$$\frac{A(z)\Delta(z)}{T(z)}\hat{y}(z) = \frac{B(z)}{T(z)}\Delta u(z) + \frac{A(z)}{T(z)}v(z) \qquad (10.6)$$

This is equivalent to process (10.2) with $F = I$ if $G(z) = A(z)^{-1}B(z)$, and $A(z)/T(z) = I$. If $A(z)/T(z) \ne I$ then prediction model (10.6) introduces bias, as for prediction one must assume $v(z)$ is unknown and therefore must be ignored.

Solving difference equation (10.6) recursively to compute the corresponding predictions (3.34) gives the prediction equation

$$\hat{y}_{k+i|k} = e_i^T[H\underrightarrow{\Delta u} + \tilde{P}\Delta\underleftarrow{\tilde{u}} + \tilde{Q}\underleftarrow{\tilde{y}}] \qquad (10.7)$$

where e_i is the ith standard basis vector and the notation is $\tilde{u} = u/T$, $\tilde{y} = y/T$. Assuming that $T(z) = A(z)$ the prediction (10.7) has no bias, so the only source of error is through the term $\tilde{Q}\underleftarrow{\tilde{y}}$. Now $\tilde{y} = \hat{y}/T$ whereas the true filtered output would be y/T. Hence the prediction error is given by

$$e_{k+i|k} = \tilde{Q}[\tilde{\underline{y}} - (y/T)] = \tilde{Q}\underline{\tilde{\zeta}}; \quad \tilde{\zeta} = \frac{\zeta}{T} \tag{10.8}$$

We use the following two assumptions to form an upper bound on the variance of this prediction error.

Assumption 1: For a zero mean variable x_k and Q a row vector, the following is true

$$E[(Q\underline{x})^2] = E[(Q_1 x_k)^2] + E[(Q_2 x_{k-1})^2] + \cdots \leq QQ^T \mathrm{var}(x) \tag{10.9}$$

Assumption 2: Consider the filter $h(z) = \sum_{i=0}^{\infty} h_i z^{-i}$ and corresponding filtered variable $\tilde{e} = he$. An upper bound on the variance of $\tilde{e}$ is given as follows:

$$\tilde{e}_k = h_0 e_k + h_1 e_{k-1} + h_2 e_{k-2} + \cdots \quad \Rightarrow \quad \mathrm{var}(\tilde{e}) \leq \mathrm{var}(e) \sum_{i=0}^{\infty} h_i^2 \tag{10.10}$$

Using these two assumptions, with $h = 1$ and $h(z) = 1/T(z)$, respectively, gives potential conservative upper bounds on error variance as follows:

- $T = 1$: The variance of $e_{k+i|k}$ in (10.8) is given from assumption 1 noting that $\mathrm{var}(\zeta) = \sigma^2$. Hence

$$\mathrm{var}(e_{k+i|k}) \leq \|\mathbf{e}_i^T Q\|_2 \sigma^2 \tag{10.11}$$

- $T \neq 1$: Using the bound (10.11) and assumption 2 (eqn. (10.10)) gives

$$\begin{aligned} \mathrm{var}(e_{k+i|k}) &\leq \|\mathbf{e}_i^T \tilde{Q}\|_2 \mathrm{var}(\tilde{\zeta}) \\ &\leq \|\mathbf{e}_i^T \tilde{Q}\|_2 \mathrm{var}(\zeta) \sum_{j=0}^{\infty} h_j^2 \\ &\leq [\|\mathbf{e}_i^T \tilde{Q}\|_2 \sum_{j=0}^{\infty} h_j^2] \sigma^2 \end{aligned} \tag{10.12}$$

Summary: One can form an upper bound on the variance of the i-step ahead prediction errors for transfer function models as

$$\mathrm{var}(e_{k+i|k}) \leq [\|\mathbf{e}_i^T \tilde{Q}\|_2 \sum_{j=0}^{\infty} h_j^2] \sigma^2 \tag{10.13}$$

10.2.1.3 Prediction and prediction errors with independent model approach

The difficulty with FIR models is that they require a large number of parameters and by necessity they must also be truncated. An independent model gives equivalent predictions, in that they are based predominantly on input information and a current disturbance estimate. However, an independent model (IM) can be constructed as a transfer function or state-space model and hence has fewer parameters and also no truncation errors.

The prediction errors will be illustrated with a transfer function model. Let the IM and associated output $\breve{y}$ be

$$A\breve{y} = Bu \tag{10.14}$$

From eqn.(3.21) predictions based on (10.14) are

$$\breve{y}_{k+i|k} = \mathbf{e}_i^T [H\Delta\underrightarrow{u} + P\Delta\underleftarrow{u} + Q\underleftarrow{\breve{y}}] \tag{10.15}$$

As with FIR model (10.3) one should correct for past disturbances and initialisation errors to ensure zero offset. Hence the predictions are adapted to

$$\hat{y}_{k+i|k} = \mathbf{e}_i^T [H\Delta\underrightarrow{u} + P\Delta\underleftarrow{u} + Q\underleftarrow{\breve{y}}] + \hat{y}_k - \breve{y}_k \tag{10.16}$$

The only error (for no parameter uncertainty) in this prediction is the use of $\hat{y}$ instead of the unknown y. Hence the prediction errors are summarised as $\text{var}(\zeta) = \sigma^2$.

Summary: The variance of the prediction errors for independent models is

$$\text{var}(e_{k+i|k}) \le \text{var}(\zeta_k) = \sigma^2 \tag{10.17}$$

This is actually better than the notional answer for FIR models due to the lack of truncation errors.

10.2.2 Analytical comparison of prediction errors from FIR, independent and transfer function models

This section contrasts the prediction errors for FIR, transfer function and IM and hence outlines a simple intuitive argument for why step response models have been so successful in practice and why transfer functions cannot be used readily without a T-filter. The discussion is based on the assumption that $F(z) = 1$ (see model (10.2) and clearly different conclusions may follow if $F(z) \ne 1$.

The models of (10.3, 10.6) are quite different in structure and these differences can have significant repercussions on prediction accuracy in cases where $v \ne 0$, $\zeta \ne 0$, see (10.5, 10.13, 10.17). These errors are now compared. To simplify the presentation (as it does not affect the conclusions) the comparisons use the assumption that the future values of v_k, ζ_k are zero. First some observations are in order.

Observation 1: It is known that the row sums of Q (see eqn.(3.21)) are all one and hence in many cases $\|\mathbf{e}_i^T \tilde{Q}\|_1 \gg 1$. Therefore in the absence of a T-filter, the prediction errors may have a much larger variance that σ^2.

A simple example illustrates this observation quite well:

$$A(z) = [1 - 0.9z^{-1}] \quad \Rightarrow \quad \mathbf{e}_1^T Q = [1.9, -0.9] \quad \Rightarrow \quad \|\mathbf{e}_i^T Q\|_2^2 = 4.42 \tag{10.18}$$

Observation 2: The case is not so clear when a T-filter is included as in eqn.(10.12). It is easy to show that $\sum(\mathbf{e}_i^T \tilde{Q}) = T(1)$ and typically $T(1) \ll 1$. However, a filter with a small bandwidth may also have $\sum_{i=0}^{\infty} h_i^2 \gg 1$.

A simple example illustrates this observation quite well:

$$T(z) = 1 - 0.8z^{-1} \quad \Rightarrow \quad \begin{cases} T(1) = 0.2 \\ \sum h_i^2 = 1/0.36 \approx 3 \end{cases} \tag{10.19}$$

So the net effect on (10.13) is hard to estimate, especially given the potential conservatism in (10.12). For instance, it is known that $T = A$ gives equivalent errors to the FIR model. Generally a low-pass T-filter reduces the error variance.

Summary: In practice one can only bound the prediction error variance for transfer function models accurately by computing the numerical values of (10.12) which implies a case by case comparison. However, simple observations indicate that the use of transfer function models may give prediction errors with a larger variance.

10.3 Numerical examples of predictions errors

This section computes the variances given in (10.5, 10.13, 10.17) and compares them with values arising from actual data.

10.3.1 Details of simulation parameters

The following two models are used for illustration:

$$\text{Example 1:} \quad G(z) = \frac{z^{-1}}{1 - 0.9z^{-1}}$$

$$\text{Example 2:} \quad G(z) = \frac{2z^{-1} + 1.4z^{-2} - 0.2z^{-3}}{1 - 1.1z^{-1} + 0.24z^{-2} - 0.05z^{-3} + 0.04z^{-4}} \tag{10.20}$$

For each of these it is assumed that $\text{var}(v_k) = \sigma^2 = 0.0393$ and the FIR models are truncated after 30 terms. The default T-filter [21] is $1 - 0.8z^{-1}$ and hence

$$\frac{1}{T} = \sum_{i=0}^{\infty} 0.8^i z^{-i}; \quad \sum_{i=0}^{\infty} h_i^2 = \frac{1}{0.36} \tag{10.21}$$

10.3.2 Summary of error comparisons

The upper bounds on prediction error variance of equations (10.5, 10.13, 10.17) are computed; these are denoted theoretical bounds. Also some open-loop simulations are performed with random inputs and the experimental prediction error variances are computed. For ease of comparison, all the variances are plotted (y-axis) against prediction horizon (x-axis). The notation used is summarised in Table 10.1. Figures 10.1–10.3 show the variance due to noise (with $v = 0$). A logarithmic sale is used because at times the difference in size of the error variances are orders of magnitude.

TABLE 10.1

Notation in Figures 10.1-10.3

	Theoretical bounds	Experimental results
Transfer function (no T-filter)	Circles	Dashed line
Transfer function (with T-filter)	Crosses	Solid line
FIR model	Squares	Dotted line
Independent model	Squares	Dash-dot line

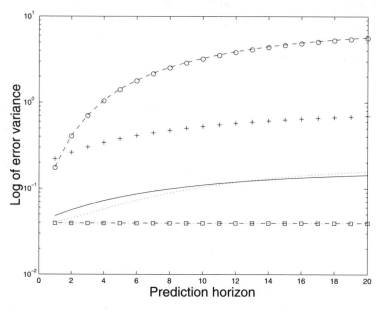

FIGURE 10.1

Variance due to noise with example 1.

10.3.2.1   Discussion of dependence of prediction errors due to noise

Figure 10.1 illustrates the results for example 1 and Figure 10.2 for example 2.

- For the transfer function without a T-filter, experimental and theoretical results match and notably are quite poor.

- The theoretical upper bounds with a T-filter are lower and the experimental results even better.

- The bounds (theoretical and experimental) for the FIR model (which is slightly worse due to truncation errors) and independent model are by far the lowest.

Figure 10.3 uses example 2 with $T(z) = A(z)$ to illustrate that the variance is then as small as with the independent model.

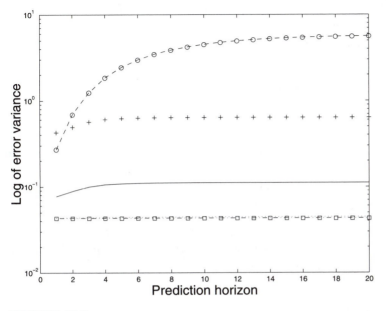

FIGURE 10.2

Variance due to noise with example 2.

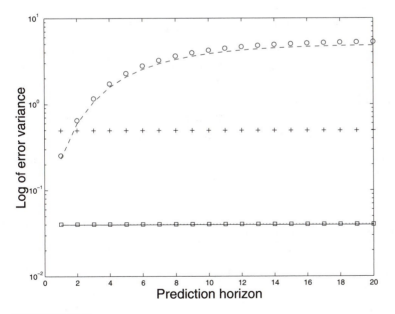

FIGURE 10.3

Variance due to noise $(T = A)$ with example 2.

Summary: For model (10.2) with $F(z) = 1$ and $v(z) = 0$, the independent model and FIR approach give the most accurate predictions. The transfer function can give equally good results if $T = A$.

10.3.2.2   Discussion of dependence of prediction errors on disturbances

The experimental results for nonzeros disturbances show a similar trend. In fact these are harder to illustrate experimentally, as an increase in error with prediction horizon is to be expected. At sampling instant k, the n-step ahead prediction cannot use information about the next n future values of v_k; hence the computed variance will increase linearly with prediction horizon, even for an ideal model. The main point to observe therefore is the difference between the strategies.

Notably the transfer function (no T-filter) is, relative to the others, very poor. Also once again the independent model often outperforms the other approaches unless $T = A$ or truncation of the FIR is not severe.

10.3.3   Conclusions

It is apparent that a simple analysis of model (10.2) with $F = 1$ gives that predictions from an FIR (or IM) model are less sensitive to noise and changing disturbances on the output. This of course is unsurprising given that the prediction is based mainly on input information.

Transfer function models realign the state at each sampling instant and hence put far more emphasis on output measurements, that is, on possibly noisy data. Incorporation of an appropriate T-filter will usually eradicate these differences, but in essence this is equivalent to using an FIR model. Moreover it is less obvious how to select $T(z)$ for the multivariable case.

In the case where there is a known disturbance model ($F(z) \neq 1$), the conclusions may be somewhat different and this needs further consideration.

Summary:

1. If you have noisy measurements, you could get noisy control action from a transfer function based MPC control law; this is unacceptable on real plant.

2. The FIR model often gives lower variance prediction errors without the need for filter (robust) design.

3. The use of an independent model is an effective means of combining low complexity models with the good performance of FIR models.

10.4 Effect of model structure on loop sensitivity

So far this chapter has looked at the impact of modelling assumptions on prediction accuracy. However, one could argue[†] that the issue of most significance is the impact on the sensitivity of the corresponding closed-loop.

The intuitive arguments will not change in that one would expect those models giving low error variances for the predictions to give rise to lower loop sensitivity to noise. Hence this section contains the following:

- A summary of the sensitivity functions for each model.

- Simulation examples contrasting the nominal loop sensitivity with different model structures.

The reader is reminded again that the complementary sensitivity is the same regardless of the model structure.

10.4.1 Control law structure and sensitivity functions

Much of the background for sensitivity functions was given in the previous chapter, so the corresponding results are simply summarised here. The exception is the control law structure that arises with an IM. It should be noted that the use of a step response and an IM model is exactly equivalent (if one assumes truncation errors are minimal) so only the results for the IM repeated here.

10.4.1.1 MPC with an independent model

This is a brief reminder of results in Section 4.6. The predictions that arise with an IM take the form of eqn.(10.16). Substitution into (4.4) and minimisation w.r.t. $\Delta \underrightarrow{u}$ gives a control law of the form (see Sections 4.3.1, 4.6)

$$\underbrace{D_k(z)\Delta\mathbf{u} = -N_k(z)\hat{\mathbf{y}} - M_k(z)\mathbf{y}; \quad \hat{A}\hat{\mathbf{y}} = \hat{B}\mathbf{u}}_{\text{Control law}} \qquad (10.22)$$

This can be simplified to

$$D_i\mathbf{u} = -M_k\mathbf{y}; \quad D_i = [D_k\Delta + N_k\hat{A}^{-1}\hat{B}] \qquad (10.23)$$

[†]An exception is the constrained case.

TABLE 10.2

Control laws for different model structures

Algorithm	Control law	Controller
GPC	$D_k \Delta u = -N_k y$	$K = [D_k \Delta]^{-1} N_k$
GPCT	$\dfrac{D_k}{T} \Delta u = -\dfrac{N_k}{T} y$	$K = [D_k \Delta]^{-1} N_k$
GPCI	$D_i u = -M_k y$	$K = D_i^{-1} M_k$

TABLE 10.3

Sensitivity to noise and disturbances

Algorithm	Output sensitivity to noise	Output sensitivity to disturbances
GPC	$S_{yn} = [A + B(D_k \Delta)^{-1} N_k]^{-1} A$	$S_{yd} = [A + B(D_k \Delta)^{-1} N_k]^{-1}$
GPCT	$S_{yn} = [A + B(D_k \Delta)^{-1} N_k]^{-1} A$	$S_{yd} = [A + B(D_k \Delta)^{-1} N_k]^{-1}$
GPCI	$S_{yn} = [A + B D_i^{-1} M_k]^{-1} A$	$S_{yd} = [A + B D_i^{-1} M_k]^{-1}$

Algorithm	Input sensitivity to noise	Input sensitivity to disturbances
GPC	$S_{un} = [D_k \Delta + N_k A^{-1} B]^{-1} N_k$	$S_{ud} = [D_k \Delta + N_k A^{-1} B]^{-1} N_k A^{-1}$
GPCT	$S_{un} = [D_k \Delta + N_k A^{-1} B]^{-1} N_k$	$S_{ud} = [D_k \Delta + N_k A^{-1} B]^{-1} N_k A^{-1}$
GPCI	$S_{un} = [D_i + M_k A^{-1} B]^{-1} M_k$	$S_{ud} = [D_i + M_k A^{-1} B]^{-1} M_k A^{-1}$

10.4.1.2   Summary of control laws

The z-transform representation of the control laws for GPC (GPC without a T-filter), GPCT (GPC with a T-filter) and GPCI (GPC based on an IM[‡]) are summarised in Table 10.2. Again, it is emphasised, as seen in Chapter 4, that D_k, N_k for GPC and GPCT will be different in general.

> **Summary:** Different model structures give different control laws and hence different loop sensitivity. However, the complementary sensitivity is the same for each.

10.4.1.3   Computation of sensitivity

For the control laws of the form given in Table 10.2 the sensitivities were fully derived in the previous chapter. The summary is given in Tables 10.3:

The sensitivity to multiplicative model uncertainty for the nominal case is

$$S_g = [I + KG]^{-1} KG \tag{10.24}$$

The different controllers to be substituted into this expression are summarised in Table 10.2.

[‡]This is equivalent to DMC.

TABLE 10.4

Notation used in plots

Algorithm	Notation
GPC	Solid line
GPCT	Dashed line
GPCI	Dotted line

10.4.2 Numerical example

The sensitivity functions for the differing model structure are best illustrated through Bode plots in the frequency range ωT is 0 to π. The notation used is given in Table 10.4.

The following example illustrates a point but *does not constitute a proof* that the same observations will follow for all examples.

For the following SISO example, the controller is designed with $n_y = 30$, $n_u = 3$, $W_u = 1$.

$$A(z) = 1 - 1.8z^{-1} + 0.81z^{-2}$$
$$B(z) = 0.01z^{-1} + 0.003z^{-2} \qquad (10.25)$$
$$T(z) = 1 - 0.8z^{-1}$$

The corresponding sensitivity functions are displayed in Figures 10.4, 10.5. A summary of the observations is given next.

- Using an IM has much reduced the input sensitivity to noise and disturbances (as well as multiplicative uncertainty) in the high frequency range.

- The output variance is also smaller for high frequencies.

- There is a larger variance of output at intermediate frequencies where one might consider noise/disturbances are less likely to occur.

- Clearly GPCT is better than GPC and more interestingly (as discussed in [158]), if $T(z) = A(z)$, the sensitivity plots of GPCT exactly replicate those of GPCI.

- GPCI has better robustness to model uncertainty.

Summary: For this example, using an IM has given good loop sensitivity without the need for a filter (or observer) design.

10.4.2.1 MIMO examples

It is also possible to plot equivalent Bode plots for MIMO examples. However, due to the interactive effects it is much harder to make intuitive observations and one would

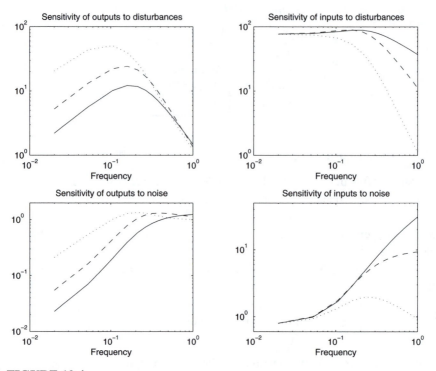

FIGURE 10.4

Closed-loop sensitivities to noise and disturbances.

need more mathematical measures to give an effective comparison. Nevertheless, for completeness the plots for a MIMO example are given next.

Consider a 2 by 2 plant with reasonably large interactions in the step response characteristics. The step responses are smooth without nonminimum phase characteristics.

$$A(z) = \begin{bmatrix} 1 & 0 \\ 0 & 1 \end{bmatrix} + \begin{bmatrix} -1.3 & 0 \\ 0 & -0.7 \end{bmatrix} z^{-1} + \begin{bmatrix} 0.4 & 0 \\ 0 & -0.18 \end{bmatrix} z^{-2} \qquad (10.26)$$

$$B(z) = \begin{bmatrix} 0.5 & 0.2 \\ -0.6 & 1 \end{bmatrix} z^{-1} + \begin{bmatrix} -0.5 & 0.3 \\ 0.3 & 1 \end{bmatrix} z^{-2} + \begin{bmatrix} 2 & 0.5 \\ 0.6 & 0.5 \end{bmatrix} z^{-3} \qquad (10.27)$$

With $T(z) = 1 - 0.8z^{-1}, n_u = 5$, $n_y = 30, W_u = 1$ the corresponding closed-loop sensitivity functions (with the notation of Table 10.4) are plotted in Figures 10.6 – 10.10, where the subplot position corresponds to the matrix position; that is, row 'i', col 'j' of the figure corresponds to $S_{i,j}$.

The observations are:

- The independent model algorithm has the lowest input and output sensitivity for high frequencies, but poorer at intermediate frequencies.

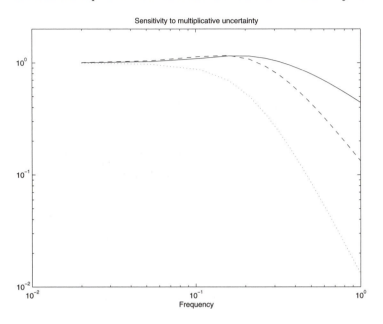

FIGURE 10.5

Closed-loop sensitivity to multiplicative uncertainty.

- For robustness to multiplicative uncertainty the case is less clear-cut though GPCI is clearly better than GPC.

Summary: The GPCI has again outperformed GPC although the MIMO characteristics make simple conclusions difficult and more mathematical measures are needed.

10.5 Conclusions

The purpose of this chapter was to highlight the issue that modelling assumptions and structure are important. Although generic conclusions are not possible, some insight into the impact of modelling assumptions on sensitivity was given. This insight could be used to improve a design before one is forced to resort to robust control theory for a more rigorous, but also more demanding, design.

Insights:

1. For the examples shown GPCI has outperformed GPC and also on average

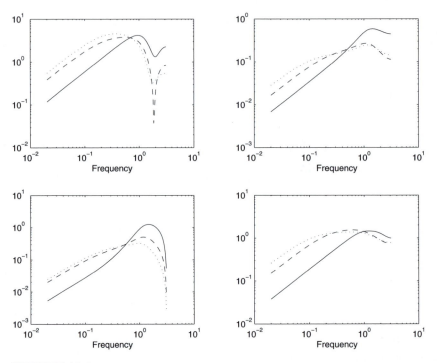

FIGURE 10.6
Output sensitivity to disturbances.

outperformed GPCT (except when $T = A$)[§].

2. GPCI often gives good sensitivity, in particular to noise, without the need to design a filter (or Youla parameter). This is not the case with GPC.

Consequent actions:

1. It is worth considering the use of an IM at the outset of a control design.

2. Sensitivity is model dependent and in practice an off-line case by case comparison is essential before the optimal model structure could be selected. The conclusions will change for different models and moreover if the disturbance model differs from that in (10.2).

One might argue, at least in the unconstrained case, that the Youla parameterisation can be used to adapt all controllers, regardless of the underlying model structure, to

---

[§]Choosing $T = A$ may not be wise or simple for the general MIMO case.

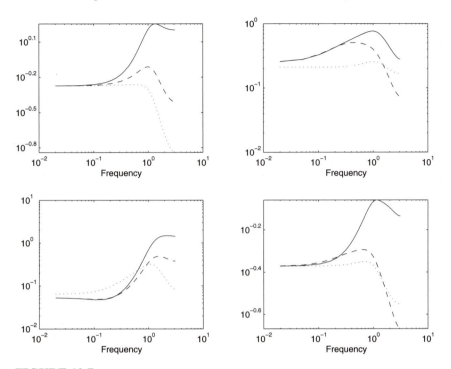

FIGURE 10.7
Input sensitivity to disturbances.

have similar robustness and to give a convenient decoupling of performance objectives from robustness objectives ([30, 60]). This is true for the constraint free case.

However, one strength of predictive control is the ability to do on-line constraint handling and the systematic extension of sensitivity functions to this case is nontrivial[¶]. In the meantime, one can argue that if the prediction structure gives low sensitivity in the nominal case, this is likely to carry over to the constrained case.

[¶]Some ideas are presented in the next chapter.

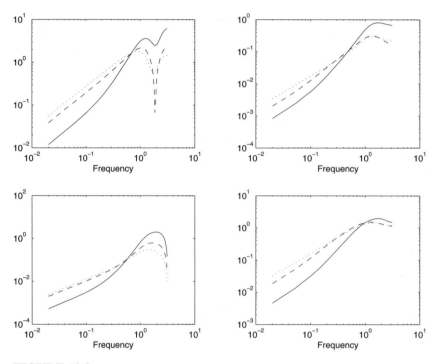

FIGURE 10.8

Output sensitivity to noise.

Summary: The reader is wise to check sensitivities before accepting a given model form (or observer).

1. The typical academic practice of using realigned models in predictive control can lead to poor sensitivity with respect to noise. A T-filter (or appropriate observer design) is usually required to overcome this but the design of a T-filter is not obvious beyond the guideline of using a low-pass filter.

2. State-space designs would need an equivalent 'robust' observer design.

3. Using an IM *often* gives low sensitivity to noise without the need for an extra design parameter. This is equivalent to a more efficient implementation of popular step response models such as used in the DMC algorithm.

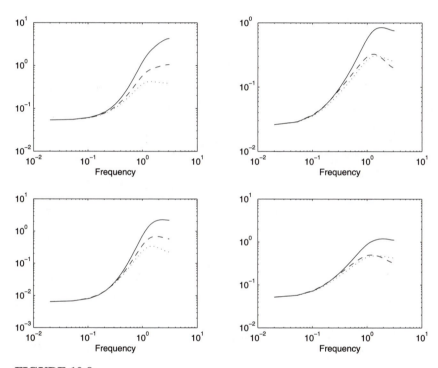

FIGURE 10.9

Input sensitivity to noise.

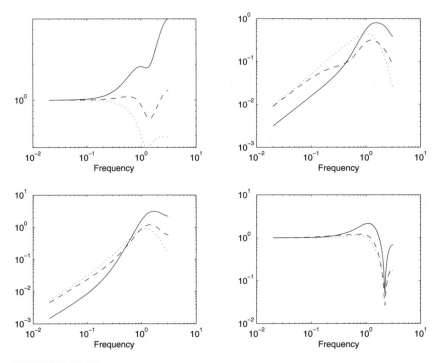

FIGURE 10.10

Sensitivity to multiplicative uncertainty.

11

Robustness of MPC during constraint handling and invariant sets

This chapter develops the work of the previous two chapters by introducing constraint handling into the uncertain case. It is separated into two major parts. The first part will look at intuitive and relatively simple ways of improving robustness and the second part will look at more rigorous, but also more computationally demanding algorithms based in invariant sets. In both approaches use will be made of the closed-loop paradigm (CLP) as it is very difficult to establish robust stability with an open-loop paradigm (OLP).

Summary: The introduction of constraints makes robustness analysis far more complex than the constraint free case.

11.1 Illustration of why robustness is hard to quantify during constraint handling

Robustness margins are difficult to determine during constraint handling because the control law becomes nonlinear and hence standard sensitivity analysis, as used in the previous few chapters, is not applicable. In fact, one could have very good sensitivity in the unconstrained case and yet the presence of a constraint could make the margin arbitrarily small.

Consider the following simple example:

$$G(z) = \frac{z^{-1}}{1 - 2z^{-1}}; \quad \bar{u} = -\underline{u} \tag{11.1}$$

Let the first input arising from an unconstrained control law be $u_0 = \bar{u}$; then it is simple to see that the process can only be stabilised if $u_i = \underline{u}, \ \forall i > 0$, that is if the input is placed on a constraint. Next add a small parameter uncertainty so that the process is actually given by $G(z) = z^{-1}/(1 - 2.0001z^{-1})$. It is now apparent that if the first control is $u_0 = \bar{u}$, the system is guaranteed unstable [61], regardless of the unconstrained stability margins; this because the system could be stabilised only by selecting $u_i < \underline{u}$ which of course is not possible.

Remark 11.1 *One could show a similar effect by adding an arbitrarily small distur-bance [40, 64].*

This problem is not noticed in the unconstrained case, as the control law is linear and hence the possible choices for u_i are not limited. Nominal sensitivity functions are applicable only when a control law is in linear operation, which implies that the state is inside the maximal admissible set. In the example given, the state has been deliberately chosen just outside the maximal admissible set (MAS) *.

It is noted that being outside the MAS does not imply instability [24] in general. More analysis is required to form the sets within which a saturated controller will give desirable behaviour.

Summary: Good sensitivity in the unconstrained case does not imply good sensi-tivity when constraints are active. The precise implications of this are difficult to generalise.

11.2   Feasibility

Feasibility is the subject of Chapter 8 but is mentioned again here, as it is a central issue when establishing robust stability in the presence of constraints.

An MPC problem is deemed feasible if it is possible, with the d.o.f. available, to satisfy all the constraints (input/output constraints and terminal constraints). One of the difficulties with establishing robustness results when constraints are present is that the answer is directly linked to feasibility. Hence in general one cannot separate the two discussions. For instance, there is little use stating that a given control law is robust with good margins if it requires infeasible control moves [†]; either the control moves will not be implementable or some other constraint will be violated.

In the robust case the problem is complicated by the fact that it maybe difficult to ensure feasibility due to the uncertainty about future behaviour. For instance, a car driver cannot guarantee to stay in lane subject to all possible actions by other drivers and the possible presence of parked cars. They could, however, give a guarantee subject to sensible assumptions on the behaviour of others. For this chapter it will be assumed that the uncertainty is bounded in some sense so that results can be derived. In general of course, uncertainty bounds are approximations and no absolute guar-antee can be given. The designer must make a judgement as to where to draw the line between the potential benefits (improved nominal performance) from ignoring

---

*The shape of the maximal admissible set will vary due to uncertainty and may not be computable in general.

[†]This is highlighted in Section 11.1.

uncertainty and the improved safety from incorporating some robust design. As yet no algorithm exists to formulate this trade-off.

Summary: Robust stability results imply a guarantee of feasibility. In practice such a guarantee requires restrictive assumptions about the uncertainty which comes at the cost of a loss in nominal performance.

11.3 Simple methods for improving robustness during constraint handling

The word *simple* here refers to the ease of coding and implementation. However, simple methods are not rigorous and hence there may be no guarantees of robust stability; rather there will be intuitive arguments for why robust performance should be improved.

11.3.1 The T-filter

The reader will recall the sensitivity improvements brought about by the introduction of a T-filter discussed in Chapter 9. The T-filter can be thought of as a low-pass filter on the output measurements and hence reduces controller sensitivity to high frequency noise but has little impact on low frequency behaviour. As this filter is hardwired into the prediction structure, the same benefits can be expected to carry across to the case where constraint handling is required (assuming feasibility). A simple illustration can be used to show this.

Take the model

$$G(z) = \frac{z^{-1}}{1 - 1.3z^{-1} + 0.4z^{-2}}; \quad |\Delta u| \leq 0.2 \tag{11.2}$$

Form a predictive controller without a T-filter (GPC, solid line) and with a T-filter $T = 1 - 0.8z^{-1}$ (GPCT, dotted line), and form closed-loop simulations in the presence of measurement noise. The unconstrained simulations are in Figure 11.1 and constrained simulations are in Figure 11.2.

The observations are:

- In the unconstrained case, GPCT clearly outperforms GPC by way of reduced input activity.

- In the constrained case, the same benefit still applies and as a consequence the ouput is also less noisy.

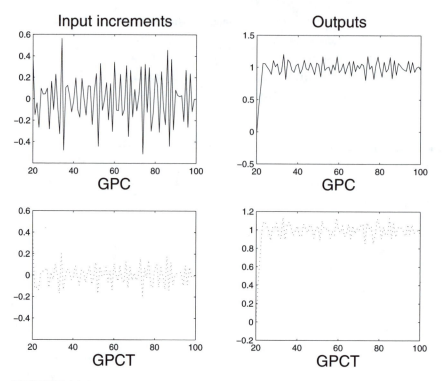

FIGURE 11.1
Closed-loop simulations without input constraints.

- If anything the improvement in performance is less for the constrained case. That is, in this case, the constraints, by limiting control movement, have actually improved the behaviour of GPC.

The effect of including the T-filter is very dependent both on the magnitude of the constraints and the noise and also on the system model. Hence it would be dangerous to take these observations too far. One cannot really generalise except to say that if the predictions (e.g. (3.21)) are less noisy, due to the filtering, then this must impact on the consequent control law, even when constraints are active.

Summary: The benefits of including a T-filter are expected to carry across to the constrained case. However, it is not possible to define analytically what the benefit will be.

11.3.2 Youla approaches

Using the same logic as above, one might think that there should be a way of using the improved sensitivity achieved via the Youla parameterisation during constraint

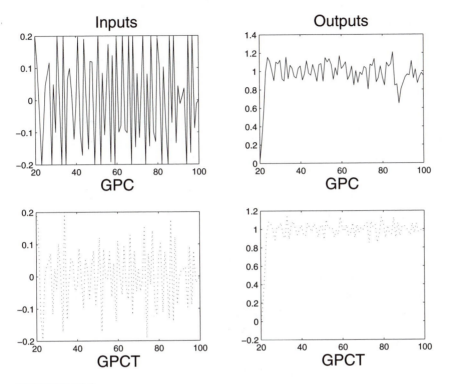

FIGURE 11.2

Closed-loop simulations with input constraints.

handling. However, there is a key difference:

- The T-filter acts on the open-loop predictions.

- The Youla approach modifies the implied closed-loop.

Hence if one is to use the Youla approach, then one must use closed-loop predictions, that is the CLP.

If one uses the CLP, then one can always design the nominal loop to be as robust as possible. The constraint handling reduces to the selection of $\underrightarrow{c}$, which does not impact on the robustness of the inner loop. However, as with the T-filter, this is equivalent to using different prediction equations (see Chapter 7). Now the prediction equations are derived from the inner loop and hence depend upon the controller parameters. The hope is that the implied feedback reduces prediction sensitivity.

Summary: Closed-loop predictions have a different sensitivity to uncertainty. Hence the CLP has the potential to be used to advantage. An illustration is given next.

11.4 Youla parameter and robust predictive control with constraint handling

This section gives a brief overview of an example presented in [114] which illustrates the potential benefits of the CLP in conjunction with a Youla parameterisation. The approach is simple to implement but does not have rigorous guarantees. The aim is to develop a means of giving a constraint handling predictive control law with a similar guarantee of robustness as achieved in the absence of constraints.

11.4.1 Nominal feedback for use in the CLP

In the CLP the set point to the loop combines with the control variable (d.o.f.) (Figure 7.1). The nominal control law (i.e. N, D) can be considered as already being chosen to both optimise performance and sensitivity, so any Youla parameter (as discussed in Section 9.4) is already incorporated. When constraints are inactive, the optimal value of c is zero and therefore the Youla parameter will increase the level of model uncertainty with which the control law can cope.

11.4.2 Introducing constraint handling

With the CLP, constraints are handled by a suitable modification of the perturbation signal **c**. However, constraint handling depends upon predictions which in turn depend upon output measurements and hence, as shown in Figure 7.1, this introduces an additional feedback loop. As with the T-filter (Section 11.3.1) the impact on sensitivity is difficult to analyse in general. One would expect that by giving the inner loop (or the closed-loop predictions themselves) better robustness properties, this is inherited to some extent by the constrained optimum.

Conjecture 11.1 *If one can prove that the perturbation signal* **c** *is convergent, then one must have robust stability.*

Proof: This is trivial as once $c = 0$, one is left with only the inner loop which is robust stable by definition.   □

The optimisation (e.g. see Section 7.5.1.2) is to minimise the weighted norm of $\underset{\rightarrow}{c}$; hence one might expect that in general **c** does converge to zero and in fact this is trivial to prove for the nominal case[‡]. Unfortunately this assumes feasibility and it is nontrivial in general to ensure that the constraints remain feasible. For this section feasibility is assumed and therein lies the weakness of the method. More rigorous

---

[‡]This follows a similar approach to that given in Section 6.3.

methods which can ensure recursive feasibility (e.g. [59, 125]) are given later in this chapter.

Summary: A stability proof requires a recursive guarantee of feasibility which is nontrivial. If one assumes recursive feasibility, then it is necessary only to show that $\underset{\rightarrow}{c}$ is stable/convergent.

11.4.3   Constraint satisfaction for a set of plants

Given that one cannot form tight equations for constraint handling over an uncertainty set, one might wonder what can be done that is simple. The most important point (as illustrated in Section 11.1) is to ensure that the choice of $\underset{\rightarrow}{c}$ is such that constraints are satisfied no matter which member of the uncertain model set is the true model.

The evolution of the inputs/outputs in the closed-loop of Figure 7.1 will vary with the model parameters, hence predicted constraint satisfaction with one assumed model G does not need to imply predicted constraint satisfaction with the model $G + \delta G$. An obvious way to avoid this pitfall, is to form (and check) the set of all possible predictions over the entire uncertainty class. However, this is not computational tractable (without using conservative bounds) in general.

Alternatively one could calculate the predictions explicitly for each of a finite set of possible plants $G^{(i)} = \frac{b^{(i)}}{a^{(i)}}$ (the superscript $(.)^{(i)}$ denotes the ith member of the model set). The input predictions, for instance, would take the form

$$
\begin{bmatrix} u_t \\ u_{t+1} \\ \vdots \end{bmatrix} = G_c^{(i)} \begin{bmatrix} c_t \\ \vdots \\ c_{t+n_c} \end{bmatrix} + \mathbf{p}_2^{(i)} \tag{11.3}
$$

where $G_c^{(i)}$, $\mathbf{p}_2^{(i)}$ depend upon N, D, $a^{(i)}$, $b^{(i)}$. One would then need to check that predictions (11.3) satisfy input constraints for all (i) (each uncertainty member), so the dimension of the constraint set is increased significantly. However, the number of d.o.f. n_c is unchanged.

Summary: If the optimising $\underset{\rightarrow}{c}$ satisfies constraints for each member of the class, then the same $\underset{\rightarrow}{c}$ must be feasible for the true model. It may not be simple in general to ensure the recursive existence of such a $\underset{\rightarrow}{c}$; in this section it is assumed.

11.4.4   Simulation study

This section will illustrate the benefits of the simple approach proposed above by way of an example. We take the example given in Section 9.4.5, which comprises a set of uncertain models and an appropriate choice for $Q(z)$. The reader is reminded

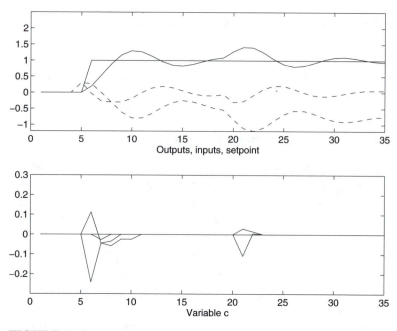

FIGURE 11.3

Perturbation δb_2, $n_Q = 0$.

that the loop is not robust stable, for the whole set of plants, without Q. Hence even in the unconstrained case, the nominal control law could fail.

Input constraints are introduced as follows:

$$|\Delta u| < 0.3; \quad -1.5 < u < 1.5 \tag{11.4}$$

Closed-loop simulations are performed where a unit set point change is demanded at the 5th sample instant and a disturbance of magnitude 0.1 enters the system at the 20th sample instant. The output, set point, input and input increment are overlaid on plots (a) and the d.o.f. $\underset{\rightarrow}{c}$ are shown on plots (b). Constraints are known to be active because $\underset{\rightarrow}{c} \neq 0$.

- Figure 11.3 shows that, without a Youla parameter and with the plant numerator given as $b + \delta b_2$, the tracking is rather bumpy.

- Figure 11.4 shows that, with a Youla parameter and with the plant numerator given as $b + \delta b_2$, the tracking is smooth, albeit with slower convergence than for the nominal plant.

- The reader is reminded that without Q, the nominal loop is unstable should the numerator be $b + \delta b_3$, but when Q is included, performance is still stable – see Figure 11.5.

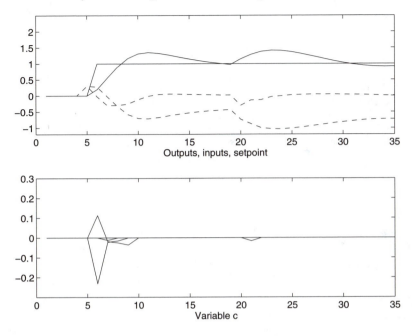

FIGURE 11.4
Perturbation δb_2 and $n_Q = 10$.

11.4.5   Conclusions

Predictive control gives a nonlinear control law when constraints are active and this implies that traditional linear robustness analysis/design cannot be implied. Nevertheless, prestabilising the plant with the nominal optimal predictive control law and using the input to this loop as the control variable, linear robust design can still be applied to some advantage.

> **Summary:** The benefits of the Youla parameterisation and formal robust design can still be applied, to some extent, during constraint handling and moreover it is a relatively simple approach to implement. However, there are no guarantees of recursive feasibility.

11.5   Using constraint tightening

This section gives a brief summary of one other simple approach (also see Section 8.5.2) that is similar in philosophy to back off and used in the process industry. If

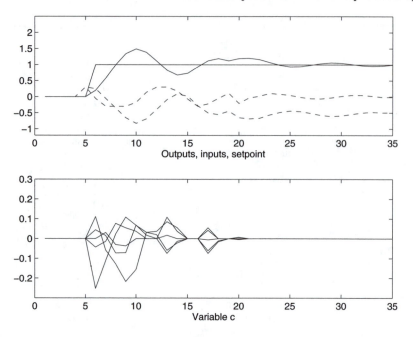

FIGURE 11.5

Perturbation δb_3 and $n_Q = 10$.

there is a danger of violating a constraint, the system is not driven as hard; that is, one backs off a little from the constraints. The amount by which one backs off is a safety margin that can be used to ensure feasibilty and stability in the event of an unexpected disturbance or other occurrence.

Unfortunately, the drawback of this approach is that guarantees, especially recursive ones, can usually only be given by deploying very conservative bounds and hence a huge sacrifice to performance. For instance, if you want to be sure not to crash your car, don't drive it! Without guarantees, the amount of back off to deploy becomes less of an analytic decision and more based on engineering judgement. As a consequence this topic is not discussed in detail in this book although the reader is reminded that a prerequisite to predicted satisfaction of constraints is that the inner loop is robustly stable; otherwise some predictions will be unstable and hence inevitably violate constraints.

It is possible [40] to find the maximum deviation between the input predictions for the nominal model a, b and the other models in the set; define these as follows

$$U_j = \max_i (u_{t+j}^{(i)} - u_{t+j}^{(0)}); \quad L_j = \min_i (u_{t+j}^{(i)} - u_{t+j}^{(0)})$$

Then one can artificially tighten the actual constraints by U_j, L_j at the upper and lower limits respectively, and this will ensure the feasibility of the whole prediction class (11.3) given the feasibility of the nominal model predictions. The advantage is that on-line one needs only to consider a single set of constraints; however, the tightening will be conservative in general.

A more detailed discussion of this approach using rigorous set theory was presented in [19].

> **Summary:** If you want to have more control freedom to deal with the unexpected, do not drive the predictions up against hard constraints except at the current sampling instant.

11.6 Recursive feasibility for the uncertain case

The results in the first half of this chapter have assumed feasibility, that is, the existence of a $\underset{\rightarrow}{c}$ such that constraints can be satisfied, at each sampling instant. For the nominal case feasibility at time k implies feasibility at time $k+1$, hence this assumption causes no difficulties. However, when there is model uncertainty and/or disturbances, feasibility does not carry over from one sample to the next in a straight-forward manner, as only one (for the true model) of the expected predictions from (11.3) actually is true and the others are all different at $k+1$ from what was conjectured at sample instant k. That is, in general (with a slight abuse of notation):

$$\underset{\rightarrow}{\mathbf{u}}_{k+1|k+1} = \underset{\rightarrow}{\mathbf{u}}_{k+1|k} \quad \not\Rightarrow \quad \underset{\rightarrow}{\mathbf{y}}_{k+1|k+1} = \underset{\rightarrow}{\mathbf{y}}_{k+1|k} \tag{11.5}$$

Exact results guaranteeing feasibility require artificially tight constraints; that is, they are conservative. Hence in practice, guarantees of feasibility for 'all possible' (most of which are highly improbable) future scenarios are not desirable due to the potential loss of performance. In practice one selects a control algorithm such that feasibility is probable, but accepts that some (unlikely) scenarios may cause infeasibility.

Nevertheless, for completeness, the remainder of this chapter looks at some algorithms which do give a guarantee of recursive feasibility and convergence for the uncertain case. The notes are deliberately brief, as this author believes that the insight gained is more important than the details of the algorithms; widespread industrial acceptance is still some way in the future.

The methods to be presented are based on a powerful paradigm which can also be used for other benefits, that is the properties of invariant sets. One might recall from Chapter 6 that dual mode MPC strategies implicitly use the property of invariance in mode 2. The remaining sections will first discuss invariant sets and second their properties. It will then be shown how invariant sets can be used to assess stability in the nominal case and then finally MPC algorithms will be proposed for establishing robust stability with constraint handling.

Summary: Guarantees of recursive feasibility in the presence of uncertainty are possible but usually require conservative assumptions and hence a degradation in nominal performance. A robust loop may be a sluggish one.

The most popular results in the literature are based on invariant sets.

11.7 Definition of invariant sets for unconstrained closed-loop systems

A set is invariant [5] if, once a state enters that set it can no longer leave. So, for instance, a set S is invariant iff:

$$\mathbf{x}_k \in \mathsf{S} \ \Rightarrow \mathbf{x}_{k+1} \in \mathsf{S} \tag{11.6}$$

The reader may like to note that condition (11.6) implies that $\mathbf{x}_{k+i} \in \mathsf{S}$, $\forall i > 0$. There are various types of invariance, e.g. [5, 56], but this book will focus mainly on controlled invariance, that is, the invariance that arises in the closed-loop system for a fixed control law.

Assuming that a system is subject to feedback, the shape of the invariant set depends upon two factors:

1. The system dynamics (or model)

2. The feedback law

In this section for ease of presentation, use will be made of state-space models and state feedback. However, invariance can equally be defined for input/output models. The underlying equations will be taken as:

1. The system dynamics: $\mathbf{x}_{k+1} = A\mathbf{x}_k = B\mathbf{u}_k$

2. The feedback law: $\mathbf{u} = -K\mathbf{x}$.

3. Closed-loop dynamics:

$$\mathbf{x}_{k+1} = \Phi\mathbf{x}_k; \quad \Phi = A - BK \tag{11.7}$$

For such a closed-loop, set S is invariant if

$$\mathbf{x}_k \in \mathsf{S} \ \Rightarrow \Phi\mathbf{x}_k \in \mathsf{S} \tag{11.8}$$

It remains now to show how sets can be constructed which satisfy this condition. Two types of invariant sets are considered here: (i) ellipsoidal sets and (ii) polyhedral sets. For simplicity of presentation it is assumed, without loss of generality, that the origin is strictly inside the invariant set.

> **Summary:** An invariant set is one which once entered cannot be left. Its definition depends upon the model and the control law.

11.7.1 Ellipsoidal invariant sets

Define an ellipsoidal set

$$S = \{\mathbf{x} : \mathbf{x}^T W \mathbf{x} \leq 1\}; \quad W > 0 \tag{11.9}$$

The closed-loop state update is given in (11.7).

Theorem 11.1 *The set S of (11.9) is invariant for model (11.7) if*

$$\Phi^T W \Phi - W \leq 0 \tag{11.10}$$

Proof: Substitute (11.7) into invariance condition (11.8), i.e.

$$
\begin{aligned}
\mathbf{x}_k^T W \mathbf{x}_k = 1 &\Rightarrow \mathbf{x}_{k+1}^T W \mathbf{x}_{k+1} \leq 1 \\
\mathbf{x}_k^T W \mathbf{x}_k = 1 &\Rightarrow \mathbf{x}_k^T \Phi^T W \Phi \mathbf{x}_k \leq 1 \\
&\Rightarrow \mathbf{x}_k^T [\Phi^T W \Phi - W] \mathbf{x}_k \leq 0, \quad \forall \mathbf{x}_k \in S \\
&\equiv \Phi^T W \Phi - W \leq 0
\end{aligned}
\tag{11.11}
$$

□

> **Summary:** An ellipsoidal set $\mathbf{x}^T W \mathbf{x} \leq 1$ is invariant for autonomous state-space model $\mathbf{x}_{k+1} = \Phi \mathbf{x}_k$ if
> $$\Phi^T W \Phi - W \leq 0$$

11.7.2 Polyhedral invariant sets

Let a polyhedral set (with no redundant constraints) be given as

$$S = \{\mathbf{x} : M\mathbf{x} - \mathbf{d} \leq 0\} \tag{11.12}$$

Theorem 11.2 *The polyhedral set (11.12) is invariant for model (11.7) if*

$$\|\mathbf{e}_i^T [M\Phi - M]\|_2 < 1, \quad \forall i \tag{11.13}$$

where $\mathbf{e}_i$ is the ith standard basis vector.

Proof: Substitution of (11.7) into condition (11.13) gives

$$M\mathbf{x} - \mathbf{d} \leq 0 \quad \Rightarrow \quad M\Phi\mathbf{x} - \mathbf{d} \leq 0 \tag{11.14}$$

Assuming (this is necessary if $\mathbf{x} = 0$ is feasible) that $\mathbf{e}_i^T \mathbf{d} > 0$, this in turn implies that

$$\mathbf{e}_i^T [M\Phi - M]\mathbf{x} < 0, \quad \forall \mathbf{x} \ \text{s.t.} \ \mathbf{e}_i^T M\mathbf{x} > 0 \tag{11.15}$$

Hence this gives a set of eigenvalue conditions which ensure that the norm is always reducing, i.e.

$$\lambda\left(\mathbf{m}_i[\Phi^T \Phi - I]\mathbf{m}_i^T\right) \leq 0 \tag{11.16}$$

$\square$

Remark 11.2 *It is clear from the above that where a matrix Φ has distinct and real eigenvalues (assumed modulus less than one as Φ is stable), one could construct an invariant set from the eigenvectors. For more general approaches the reader is referred to the literature, e.g. [5].*

Summary: A polyhedral set $\mathsf{S} = \{\mathbf{x} : M\mathbf{x} - \mathbf{d} \leq 0\}$ is invariant for autonomous model $\mathbf{x}_{k+1} = \Phi\mathbf{x}_k$ if $\lambda\left(\mathbf{m}_i[\Phi^T \Phi - I]\mathbf{m}_i^T\right) \leq 0, \ \forall i$.

11.7.3 Link between invariance and stability

The reason why invariance is so popular as a tool is that an invariance condition is equivalent to a stability test. For instance, one could take the invariance test as equivalent to a Lyapunov function.

- The ellipsoidal invariant set (11.9) could be considered to represent a Lyapunov function $J = \mathbf{x}^T W \mathbf{x}$. If the invariance condition is satisfied, then J is monotonically decreasing. As W is positive definite, this implies that $\mathbf{x}$ is convergent.

- The polyhedral invariant set (11.12) could also be considered to represent a Lyapunonv function $J = \|M\mathbf{x}\|_\infty$. Again if condition (11.16) is satisfied, then J must be monotonically decreasing which in turn implies that $\mathbf{x}$ must be convergent.

Summary: The existence of an invariant set is equivalent to the existence of a Lyapunonv function and hence is equivalent to a stability test.

11.8   Invariance and constraint handling

It has just been shown that the existence of an invariant set is equivalent to a Lyapunov stability test. As was noted in Chapter 6 Lyapunov stability tests can also be applied during constrained handling. This section will show how invariant sets can be used to guarantee convergence in the presence of constraints.

11.8.1   Ensuring constraint satisfaction by set membership

A few lemmata are used here to show how one can use a set membership test to ensure predicted constraint satisfaction. This section deals only with the principle that the sets exist and does not tackle the question of how the invariant sets should be computed.

Lemma 11.1 *If there exists an invariant set (11.8) for autonomous model (11.7), then this set can always be scaled to be small enough so that constraints are always satisfied.*

Proof: The input is given by the feedback $\mathbf{u} = -K\mathbf{x}$ and $\mathbf{x}$ is restricted to satisfy $\mathbf{x}^T W \mathbf{x} \leq 1$. Hence [59], we can always take W large enough, so that the allowable values for $\mathbf{x}$ ensure $K\mathbf{x}$ is smaller than the input limits.   □

Lemma 11.2 *If there exists an invariant set (11.12) for autonomous model (11.7), then this set can always be scaled to be small enough so that constraints are always satisfied.*

Proof: The constraints can be stated naturally as linear inequalities, for instance, $u < \bar{u} \; \rightarrow \; -K\mathbf{x} \leq \bar{u}$. Hence one needs only scale the rows of M to be large enough so that $M\mathbf{x} - \mathbf{d} \leq 0$ implies that $\mathbf{x}$ is small enough that $-K\mathbf{x} \leq \bar{u}$. In general of course, one would assume that constraint $-K\mathbf{x} \leq \bar{u}$ was included systematically in the choice of M, $\mathbf{d}$ (recall that the choice of K is implicit in the definition of M anyway).   □

> **Summary:** If there exists an invariant set for an autonomous model, then one can find an invariant set within which the closed-loop trajectories always satisfy constraints.

11.8.2   Using invariant sets in predictive control

One major ability of MPC is on-line constraint handling, that is, the ability to optimise performance subject to there being no predicted constraint violations. However, in practice the predictions evolve over an infinite horizon, especially in the case of dual mode algorithms and this seems to imply that the constraints should be checked over an infinite number of sampling instants. Invariant sets are the key to overcoming this obstacle.

First remind yourself of the two key points:

- Once a state enters an invariant set, it never leaves.

- An invariant set can be defined so that if the state is within it, then the state/input trajectories of the closed-loop system do not violate any constraints.

Now consider the dual mode paradigm (see Section 6.6.3). The terminal mode of a dual mode prediction is equivalent to unconstrained behaviour of a known closed-loop. Hence this behaviour can be captured by an invariant set and more importantly we can make the following lemma.

Lemma 11.3 *Let* S *be an invariant set for system A and control law B and furthermore let this set be scaled such that the closed-loop input/state trajectories satisfy constraints C. Then constraint satisfaction in mode 2 is ensured by membership of* S *.*

Proof: The proof is obvious though one should note that membership of S is sufficient but may not be necessary. □

Theorem 11.3 *The constraints used in a dual mode algorithm can be posed as: (i) an explicit comparison of input/state predictions with their respective limits during mode 1 and (ii) the membership of an appropriate invariant set as the state enters mode 2.*

Proof: Again this is obvious. □

Remark 11.3 *The most notable thing about this theorem is the following observation. The invariant set will be finitely determined. Hence constraint satisfaction over an infinite horizon can be ensured by checking a finite number of inequalities:*

- *Linear inequalities representing constraints during mode 1.*

- *Either linear inequalities (11.12) or quadratic inequalities (11.8) ensure membership of the terminal invariant set.*

Summary: Membership of an appropriate invariant set is equivalent to testing for constraint satisfaction of the closed-loop predictions over an infinite horizon. The number of inequalities in the implied invariant set will be finite and often quite small.

11.9 Computing invariant sets

It remains now to show how invariant sets can be computed in general. We note, as discussed the Chapter 8 and was indicated in Section 7.4.1, that there is a general desire for the terminal invariant sets to be as large as possible, as this reduces conservatism and hence suboptimality. However, the desire to maximise volume may come into conflict with the desire for a simple definition.

This section focuses first on computing invariant sets for the certain case. The uncertain case is discussed in a subsequent section.

11.9.1 Computing polyhedral invariant sets

The MAS was developed in [33]. This set is the largest possible invariant set for a given system, constraints and control law and in that sense membership of the set is both necessary and sufficient for constraint satisfaction assuming the specified linear control. This follows from the fact that any states outside the set, by definition, are violating a constraint. In the following an overview is given of how to determine the MAS.

Given that $\mathbf{u} = -K\mathbf{x}$, let the constraints (input/output/state) at each sampling instant be represented by the set of linear inequalities:

$$C\mathbf{x}_{k+i} - \mathbf{d} \leq 0, \ \ \forall i \geq 0 \tag{11.17}$$

Let the closed-loop dynamics be $\mathbf{x}_{k+1} = \Phi\mathbf{x}_k$ and define the candidate set $\tilde{S}_n$ in which constraints are satisfied for the first n steps of the predicted response as

$$\tilde{S}_n = \{\mathbf{x} : \underbrace{\begin{bmatrix} C\Phi \\ C\Phi^2 \\ \vdots \\ C\Phi^n \end{bmatrix}}_{C_n} \mathbf{x} - \underbrace{\begin{bmatrix} \mathbf{d} \\ \mathbf{d} \\ \vdots \\ \mathbf{d} \end{bmatrix}}_{\mathbf{d}_n} \leq 0\} \tag{11.18}$$

Set $\tilde{S}_n$ is invariant if $\mathbf{x} \in \tilde{S}_n \ \Rightarrow \ \Phi\mathbf{x} \in \tilde{S}_n$. The test for invariance is by contradiction; i.e. does there exist $\mathbf{x} \in \tilde{S}_n$ such that $\Phi\mathbf{x} \notin \tilde{S}_n$. More specifically, one can maximise (w.r.t. $\mathbf{x}$) each of the row sums of $C_n x - \mathbf{d}_n$, in turn subject to the remainder of the constraints in $\tilde{S}_n$ being satisfied. If any of the maximal row sums is greater than zero, then there exists an $\mathbf{x}$ such that $\mathbf{x} \in \tilde{S}_n$ but $\Phi\mathbf{x} \notin \tilde{S}_n$.

Hence the formulation of the MAS is via a sequence of linear programs or iteration:

1. Define $\tilde{S}_1$, set $i = 1$.

2. Test whether $\tilde{S}_i$ is invariant.

3. If $\tilde{S}_i$ invariant, exit else $i = i + 1$ and go to step 2.

Convergence: This iteration will terminate with finite i so long as the asymptotic point for the state; that is, $\lim_{i \to \infty} \mathbf{x}_{k+i} = \lim_{i \to \infty} \Phi^i \mathbf{x}_k$ is strictly inside the interior of the constraints (11.17).

Observation: Many of the rows in the MAS $\tilde{S}_n$ may be redundant and hence it is often worthwhile removing these.

Remark 11.4 *There has been some work on limited complexity invariant polyhedrals using conditions such as (11.16), but that work is large in itself and hence is not discussed here. The motivation of the work is to derive an invariant set defined by far fewer inequalities. Of course the price one pays is that the set will be conservative in volume; that is, membership is a sufficient but not necessary test for feasibility.*

Summary: Given constraints (11.17) at each sampling instant, the MAS is invariant and defined as

$$\tilde{S}_n = \{\mathbf{x} : \begin{bmatrix} C\Phi \\ C\Phi^2 \\ \vdots \\ C\Phi^n \end{bmatrix} \mathbf{x} - \begin{bmatrix} \mathbf{d} \\ \mathbf{d} \\ \vdots \\ \mathbf{d} \end{bmatrix} \leq 0\} \qquad (11.19)$$

for large enough n. For typical closed-loop poles, say modulus 0.8 to 0.9, n is likely to be in the region 10–20 and hence may not be large.

11.9.2   Ellipsoidal sets

Ellipsoidal invariant sets have the advantage of having a simple definition (11.8) which reduces complexity, but they have the disadvantages of being both suboptimal in volume (as the MAS is known to be polyhedral) and more difficult to compute. Moreover, using ellipsoidal terminal sets within MPC implies the mixing of linear and quadratic constraints in the MPC optimisation (e.g. modify the optimisation in Section 7.2.3 by replacing the MAS with constraint (11.8)) and this can lead to a nontrivial calculation (e.g. [67]).

Nevertheless, with some modification of objectives, ellipsoidal sets can also be used to computational advantage [65] and to handle robust problems [59]. Hence this section outlines some methods by which ellipsoidal invariant sets can be computed.

11.9.2.1   Simple choices of ellipsoidal invariant set

A simple choice of invariant set arises from the level set of the performance index (for the infinite horizon case). It was shown in Section 6.6 that the unconstrained optimum could be written as $J = \mathbf{x}^T P \mathbf{x}$ and moreover that this was a Lyapunonv function. Hence P is a suitable candidate for the matrix W of (11.8).

In an equally simple fashion, one could make use of the eigenvalue/vector decomposition of Φ and the observation that the eigenvalues are all modulus less than one to form a suitable ellipsoid.

Both these choices may be quite conservative in volume by comparison with other ellipsoids and hence are not favoured in general.

11.9.2.2 Maximal volume ellipsoidal sets

The condition for invariance is given as

$$\Phi^T W \Phi - W < 0 \tag{11.20}$$

This can be represented as a linear matrix ineqaulity(LMI) [11, 59].

$$\begin{bmatrix} W^{-1} & \Phi W^{-1} \\ W^{-1} \Phi^T & W^{-1} \end{bmatrix} \geq 0 \tag{11.21}$$

Input constraints can be handled using the following observation:

$$\begin{aligned} |K_i^T \mathbf{x}|^2 &\leq |K_i^T W^{1/2} W^{-1/2} \mathbf{x}|^2 \leq \|K_i^T W^{1/2}\|_2^2 |W^{-1/2} \mathbf{x}|^2 \\ &\leq (K_i^T W K_i)(\mathbf{x}^T W^{-1} \mathbf{x}) \leq K_i^T W K_i \leq \bar{u}^2 \end{aligned} \tag{11.22}$$

Hence the constraints $-K_i^T \mathbf{x} \leq \bar{u}$ could be achieved via satisfaction of the LMIs

$$\begin{bmatrix} W^{-1} & W^{-1} K_i \\ K_i^T W^{-1} & \bar{u}^2 \end{bmatrix} \geq 0; \quad i = 1, 2, \dots \tag{11.23}$$

Finally, the set given in (11.8) is invariant and moreover constraints are satisfied if both LMIs (11.21, 11.23) are satisfied and $W > 0$ (positive definite). Recall however, that these conditions are sufficient but not necessary.

Remark 11.5 *Lower constraints on the input and state constraints give rise to LMIs similar to (11.23). Hence each separate constraint will give rise to an additional LMI to be satisfied. Other constraints can also be handled in a similar fashion.*

Theorem 11.4 *The maximum volume invariant elliposid such that constraints are guaranteed to be satisfied can be computed from the following optimisation:*

$$Max \log \det(W^{-1}) \quad \text{s.t.} \quad (11.21, 11.23) \tag{11.24}$$

Proof: LMI (11.21) ensures invariance. The set of LMIs in (11.23) ensure constraint satisfaction inside the set. The volume of an ellipsoid is inversely proportional to the product of the eigenvalues, that is, the determinant. □

Remark 11.6 *LMI methods are becoming popular in the literature but still require quite significant computation compared, for instance, to a conventional quadratic programming (QP) optimisation. The real potential is in the application to model uncertainty and nonlinearity. There is not space in this book to discuss this properly and so the reader is referred elsewhere for a detailed study of LMI techniques.*

Summary:

- An ellipsoidal invariant set within which constraints are satisfied is given by $\mathbf{x}^T W \mathbf{x} \leq 1$ where conditions (11.21, 11.23) and $W > 0$ all apply.

- For the certain case the ellipsoidal set is suboptimal in volume and hence its use may unnecessarily restrict the regions within which an MPC algorithm is defined.

11.10   Invariance in the presence of uncertainty

One important motivation for introducing invariant sets was to handle uncertainty; this section demonstrates how that can be achieved. First MAS are discussed and dismissed and then some space is given to ellipsoidal sets.

In practice all systems exhibit some uncertainty, by way of disturbances or parameter uncertainty. Clearly the invariance conditions (11.6) or (11.16) may no longer be valid in the presence of uncertainty and the conditions need reformulating.

Uncertainty affects the autonomous model assumption whereby (11.7) must be replaced by:

1. Disturbance uncertainty:

$$\mathbf{x}_{k+1} = \Phi \mathbf{x}_k + \beta \qquad (11.25)$$

   where β is unknown but possibly bounded.

2. Parameter uncertainty: $\mathbf{x}_{k+1} = [\Phi + \Delta_\Phi]\mathbf{x}_k$

For convenience hereafter, we will quantify the parameter uncertainty using linear differential inclusions; for instance, let the closed-loop state-space matrix be described as:

$$\Phi = \sum \mu_i \Phi_i, \quad \sum \mu_i = 1, \quad \mu_i \geq 0 \qquad (11.26)$$

Summary: In the presence of uncertainty, the invariance conditions need reformulating so that they apply to the whole uncertainty class. To do this an uncertainty class must be defined.

11.10.1   Polyhedral sets and uncertainty

The MAS defined in (11.19) was for the nominal case only. Yet even here the complexity could be considered quite large. If instead one were to use model (11.26), then to check and allow recursive constraint satisfaction over a prediction horizon of *just 2*, one would require the following linear inequalities:

$$
S = \{\mathbf{x} : \begin{bmatrix} C\Phi_1 \\ \cdots \\ C\Phi_n \\ C\Phi_1\Phi_1 \\ C\Phi_1\Phi_2 \\ \vdots \\ C\Phi_n\Phi_{n-1} \end{bmatrix} \mathbf{x} - \begin{bmatrix} \mathbf{d} \\ \mathbf{d} \\ \vdots \\ \mathbf{d} \end{bmatrix} \leq 0\}
\tag{11.27}
$$

Evidently there is a combinatorial explosion in the number of inequalities which is simply not manageable for large prediction horizons.

Hence, in general, one cannot form an MAS in the uncertain case. In fact a more fruitful way forward is to return to the definition given in (11.16), predefine the complexity (number of inequalities) and then search for appropriate coefficients [5]. This avenue is not pursued in this book due to limitations in space and a desire not to get into more mathematical algorithms.

> **Summary** The MAS is not easy to define in the uncertain case. Algorithms do exist which search for low dimensional invariant polyhedrals to cater for uncertainty, but those are outside of the scope of this book.

11.10.2   Ellipsoidal invariance in the presence of uncertainty

The topic of ellipsoidal invariance will also not be discussed in detail here as the mathematics quickly gets more complex than fits in the scope of this book. Hence only the main concepts are outlined. Two issues are discussed separately: (i) uncertainty due to exogeneous signals such as disturbances and (ii) parameter uncertainty. More generally one can also consider mild nonlinearity but that is not discussed here.

11.10.2.1   Disturbance uncertainty

The most important observation [19] here is that, perhaps counter to one's intuition, in the presence of disturbances small invariant sets are not possible. This in turn means that for either a well tuned loop or one with tight constraints it may not be possible to define an invariant set.

The explanation is simple. The invariance condition can be written as

$$\mathbf{x}_{k+1}^T W \mathbf{x}_{k+1} \le \mathbf{x}_k^T W \mathbf{x}_k \tag{11.28}$$

$$[\mathbf{x}_k^T \Phi^T + \beta^T] W [\Phi \mathbf{x}_{k+1} + \beta] \le \mathbf{x}_k^T W \mathbf{x}_k$$
$$\mathbf{x}_k^T \Phi^T W \Phi \mathbf{x}_{k+1} + 2\beta^T W \Phi \mathbf{x} + \beta^T \beta \le \mathbf{x}_k^T W \mathbf{x}_k$$
$$\mathbf{x}_k^T \Phi^T W \Phi \mathbf{x}_{k+1} \le \mathbf{x}_k^T W \mathbf{x}_k - 2\beta^T W \Phi \mathbf{x} - \beta^T \beta$$

Now consider the case where $\mathbf{x}$ is small (or even zero), then the invariance condition reduces to

$$0 \le 0 - \beta^T \beta \tag{11.29}$$

and clearly this is inconsistent.

Well tuned controllers in conjunction with input constraints often result in small invariant sets so one can quickly set up a contradiction such that for certain input limits in the face of uncertainty, an invariant set will not exist for a given fixed state feedback. In the presence of disturbance uncertainty, the invariant set must be big enough so that for a state $\mathbf{x}$ on the boundary, then $\mathbf{x}^T[W - \Phi^T W \Phi]\mathbf{x} > \beta^T \beta$. Hence the larger the possible disturbance signal, the larger the invariant set needs to be.

It is quite possible that the size of invariant set required comes into conflict with the LMI requirements of Theorem 11.4, in particular constraints (11.23); and then a simple invariant set cannot be defined.

This is still an active research area and the material is not really suitable for this book. One might conjecture however, that seeking such rigor in a real world scenario is perhaps an unrealistic objective.

Summary: In the presence of disturbances and constraints, it maybe impossible to define an invariant set. This is because invariance requires the natural *change* in the state always to be larger than the effect of the disturbance.

11.10.2.2   Parameter uncertainty

This case can be handled more easily than the above because parameter uncertainty is proportional to the magnitude of the state, whereas a disturbance signal is not. Hence one can obtain realistic invariant sets.

Consider the condition (11.20) for invariance for a certain process. This must be satisfied for each member of the class of uncertainty, that is:

$$\Phi_i^T W \Phi_i - W < 0; \quad \forall i \tag{11.30}$$

This gives rise to a number of LMIs conditions analogous to (11.21).

Summary: One can easily state the LMI conditions for an invariant ellipsoid to exist in the case of parameter uncertainty. However, this does not need to imply the conditions can be satisfied or that the implied computation is simple.

11.11 Using ellipsoidal invariant sets in robust MPC design

This chapter will only mention algorithms for robust design in passing as I believe the technology is still several years ahead of implementation. However, it is useful to be aware of them, as recent work on robust MPC has almost universally been based on invariant sets and LMIs. There exist [11] many efficient tools for solving LMI problems, although these are still slower than QP. Also LMI methods can be used for off-line analysis, to give confidence in expected algorithm behaviour.

The power of invariant sets is that if one can ensure that the predictions enter an invariant set, then once inside stability/convergence/constraint satisfaction are guaranteed. Moreover, as noted in Section 11.10.2.2 the set can be defined for an uncertain class; that is, it is not restricted only to the nominal case. The difficulty is then simplified to finding a robust method of driving the state inside an invariant set (which is naturally larger than the origin).

11.11.1 Overview

There are two main branches to the use of LMIs and ellipsoidal invariance:

1. The first branch [59] allowed the feedback K to be the d.o.f. and searched for a K such that the current state was inside an invariant set for the given constraints and model uncertainty. This could give rise to cautious control and requires a significant on-line computation.

2. The second approach predetermined the K and found the maximum volume invariant set (for the given number of d.o.f.) [66]. This could give better performance and required only a small on-line computation [65] but feasibility was restricted by the off-line assumption on the underlying K.

3. Most results are still restricted to ellipsoidal[§] spaces (but see [15]) and this is a severe restriction given that realistic MAS are polyhedral.

11.11.2 Algorithm of Kothare et *al*

In [59] the authors took the premise of a linear time varying (LTV) process where the state matrices $A(k), B(k)$ lay somewhere inside a polytope. However, its time variation was assumed unknown. Let the vertices of such a polytope be given by A_i, B_i, $i = 1, \ldots$ In order to guarantee convergence the objective was to find a state

---

[§]Or suboptimal polyhedral spaces.

feedback (full state knowledge was assumed) such that

$$|\lambda(A_i - B_iK)| < 1, \quad \forall i \tag{11.31}$$

Given that $\Phi_i = A_i - B_iK$, choosing K is nontrivial so the problem was replaced by an alternative test. Define an ellipse as

$$S = \{\mathbf{x} : \mathbf{x}^T P \mathbf{x} \leq 1\} \tag{11.32}$$

where P is yet to be selected. Now S is invariant if $\mathbf{x}_k \in S \Rightarrow \mathbf{x}_{k+1} \in S$ for each model in the set. Therefore the test for stability of the uncertain system is equivalent to the test

$$\Phi_i^T P \Phi_i - P < 0, \quad \forall i; \quad P > 0 \tag{11.33}$$

The objective then is to parameterise K in such a way that one can optimise the predicted performance subject to inequalities (11.33) and subject to $\mathbf{u} = -K\mathbf{x}$ satisfying input constraints (see eqn. 11.23). The d.o.f. in the optimisation are both K and P, as the only requirement on P is that it is positive definite.

One then finds that the optimal K is time varying, so as the state moves nearer the origin K becomes gradually more highly tuned. The downsides of this algorithm are that:

1. The computation of K arises from a very involved LMI computation, at each sampling instant.

2. At each step, it is assumed that in the predictions the control law is linear. However, it is known that, in general, the optimal trajectory (during constraint handling) is nonlinear during the transients.

There have been many subsequent developments in the literature.

Summary: One can formulate an MPC algorithm that handles constraints with a guarantee of recursive feasibility and convergence and also allows for parameter uncertainty. However, the on-line computational load is large and the associated theory is demanding. The algorithm is restricted to ellipsoidal regions.

11.11.3   Using the closed-loop paradigm

It was shown in [66] that one could build on the above algorithm and derive a far simpler algorithm by deploying the CLP. The key idea is to use transient d.o.f. to enlarge the volume of the invariant set rather than changes in the underlying control law.

The algorithm proposed transferred the major computational burden from on-line to off-line and also simplified the off-line computation to the search for ellipsiodal sets only (the implied K is given unlike in [59]).

Off-line:

- Choose a nominal control law (with optimised robustness) and find an ellipsoidal invariant set. (The volume/shape depends on constraints.)

- Assume that pertubations $\underset{\rightarrow}{c}$ will be used. Hence create an autonomous system in which a finite number of future c appear as states.

- Find the maximum volume ellipsoidal sets for the augmented system.

On-line:
Select the minimum norm $\underset{\rightarrow}{c}$ so that the augmented state is inside the appropriate ellipsiodal set. The computation reduces to finding the only positive real root of a polynomial. This is trivial and in fact far simpler than a QP [65].

> **Summary:** If you are prepared to fix some parameter, such as the underlying control law, then one can simplify the algorithm while still handling the robust case. However, the feasible region is limited by the assumptions made.

11.11.3.1 Summary

It is re-emphasised that the robust algorithms make use of invariant sets and as such the algorithms are only defined when the state is within those sets. However, there may be many points outside the invariant sets [56] for which a robust control strategy exists. Hence the algorithms can still be very conservative.

> **Summary:** To obtain a guarantee of robust stability in the presence of constraints, it is likely that the associated algorithm will give conservative performance and be valid only within a quite restricted region.

11.12 Conclusions

This chapter has shown that invariant sets are invaluable for improving performance of predictive control, extending applicability and also allowing analysis of expected behaviour. The key message is that set membership is equivalent to testing for constraint satisfaction, over an infinite horizon and possibly for the uncertain case. A known control law is implied within the set definition and hence set membership also gives a handle on the performance.

Summary:

1. MPC algorithms are often defined as optimising some cost (usually with infinite output horizons) subject to constraints during the first n_c samples and membership of an invariant set within n_c steps. This automatically gives recursive feasibility.

2. The definition of the invariant set is a major tuning parameter for small n_c, as it carries an implied control law from the n_c^{th} step to the infinite horizon (see Section 7.3).

3. The use of invariant sets in MPC allows relatively straightforward extension to the uncertain case.

12

Optimisation and computational efficiency in predictive control

One of the key components of an MPC algorithm is the optimisation. However, optimisation is such a large topic in itself that it would be difficult to say much meaningful in this book. Hence instead the focus will be to:

- Indicate the main optimisation algorithms used by MPC and give some brief discussion upon them.

- Look at how the optimisation burden can be reduced by changes in objective.

Issues of how to improve the robustness or efficiency of an optimisation algorithm itself are considered to be outside the scope of the control engineer, although of course they are important. Some discussion of these points appears in [13, 78] and hence is not repeated here.

Predictive control strategies allow for the systematic handling of constraint, performance, and stability. However, the associated algorithms can be computational burdensome and/or difficult to unravel. This chapter will discuss and compare a few algorithms based on invariant sets which meet the additional requirement for computational simplicity. There may of course be a concomitant loss of optimality, but this can be minimal and often is a small price to pay when one considers the significant improvements in efficiency.

Summary: This chapter focuses mainly on how the MPC algorithm itself can be modified to reduce computational load. For a full discussion of optimisation algorithms, the reader is referred elsewhere.

12.1   Optimisation algorithms in MPC

It will be apparent (e.g. Section 4.8.2) by now that the most typical MPC algorithms require the on-line solution of a quadratic programming (QP) problem. For GPC *

---

*It is assumed that readers can supply their own details for CLP and dual mode algorithms.

this can be summarised as:

$$\min_{\underset{\rightarrow}{\Delta u}} \quad J = \Delta \underset{\rightarrow}{u}^T S \Delta \underset{\rightarrow}{u} + 2 \Delta \underset{\rightarrow}{u}^T f \quad \text{s.t.} \quad C \Delta \underset{\rightarrow}{u} - d \le 0 \tag{12.1}$$

The question that may be on the reader's mind is, *how easy is a QP to solve?*

12.1.1 Active set methods

The most common method for solving a QP is the active set method (ASM). This book will not give fine details as the interested person should read a book on optimisation. The QP algorithm is available within MATLAB and hence readily accessible to researchers as a tool.

At the constrained optimum, some of the constraints would be active and the remainder would be inactive. The aim of the ASM is to find this separation. Here a brief summary of the key steps is given.

1. Ignoring ordering issues for simplicity, separate the constraints as follows[†]:

$$\left. \begin{array}{l} C_{ac} \Delta \underset{\rightarrow}{u} - d_{ac} = 0 \\ C_{in} \Delta \underset{\rightarrow}{u} - d_{in} < 0 \end{array} \right\} ; \quad C = \begin{bmatrix} C_{ac} \\ C_{in} \end{bmatrix} ; \quad d = \begin{bmatrix} d_{ac} \\ d_{in} \end{bmatrix} \tag{12.2}$$

2. Then, ignoring the inactive constraints $C_{in} \Delta \underset{\rightarrow}{u} - d_{in} < 0$, one can rewrite the optimisation (12.1) as

$$\Delta \underset{\rightarrow \text{test}}{u} = \arg \min_{\underset{\rightarrow}{\Delta u}} \quad J = \Delta \underset{\rightarrow}{u}^T S \Delta \underset{\rightarrow}{u} + 2 \Delta \underset{\rightarrow}{u}^T f \quad \text{s.t.} \quad C_{ac} \Delta \underset{\rightarrow}{u} - d_{ac} = 0 \tag{12.3}$$

3. Substitute $\Delta \underset{\rightarrow \text{test}}{u}$ into the inactive constraints and compute l

$$l = \max(C_{in} \Delta \underset{\rightarrow \text{test}}{u} - d_{in}) \tag{12.4}$$

   - If $l \le 0$, then $\Delta \underset{\rightarrow \text{test}}{u}$ is a feasible solution and may be the optimum. One can test whether $\Delta \underset{\rightarrow \text{test}}{u}$ is the global optimum by looking at the associated lagrange multipliers, which should be positive.
   - If $l > 0$, then $\Delta \underset{\rightarrow \text{test}}{u}$ is infeasible and cannot be the optimal.

4. If $\Delta \underset{\rightarrow \text{test}}{u}$ is not the global optimum, change[‡] the definition of active and inactive sets and redo steps 1 to 3.

Summary: The ASM gives a systematic means of selecting a potential active set and iterating through these potential sets to find the global optimum. Although the upper limit on the number of iterations is prohibitively large, it is rarely approached in practice and the ASM is widely used.

[†]Typically one could initialise by setting as active those constraints violated by the unconstrained optimum.

[‡]The details of how this update occurs is outside the scope of this book.

12.1.2 Interior point methods

These are becoming more popular than ASM methods within MPC, as the convergence rates are far faster. However, the associated optimisation at each iteration is more demanding. Again the reader is referred elsewhere (e.g. [93, 155]) for a detailed discussion and only a brief summaryis given here.

1. Replace the optimisation (12.1) by the optimisation (e.g. see [101] for an early variant):

$$\min_{\underset{\rightarrow}{\Delta\mathbf{u}}} \quad J = \underset{\rightarrow}{\Delta\mathbf{u}}^T S \underset{\rightarrow}{\Delta\mathbf{u}} + 2\underset{\rightarrow}{\Delta\mathbf{u}}^T \mathbf{f} + \|W^{(i)}(C\underset{\rightarrow}{\Delta\mathbf{u}} - \mathbf{d})\| \tag{12.5}$$

where $W^{(i)}$ maybe a diagonal weighting matrix at the ith iteration or a function.

2. $W^{(i)}(C\underset{\rightarrow}{\Delta\mathbf{u}} - \mathbf{d})$ is chosen to approach infinity[§] as the constraint becomes active, and so the optimum for (12.5) will always choose $C\underset{\rightarrow}{\Delta\mathbf{u}} - \mathbf{d} < 0$; hence we have the name interior point.

> **Summary:** Interior point methods are guaranteed to converge, within a given accuracy, much faster than QP algorithms. Hence these are becoming popular.

12.1.3 Multi parametric quadratic programming (MPQP)

This is a relatively recent development (e.g. [7, 8]) which is still an area of active research. The algorithm makes use of the observation in Section 12.1.1 that, should a given active set be feasible, the optimisation (12.1) can be replaced by (12.3). As this is a quadratic program with equality constraints, then the solution has a fixed form, that is one could write:

$$\underset{\rightarrow test}{\Delta\mathbf{u}} = -K_{ac}\mathbf{x} + \mathbf{k}_{ac} \tag{12.6}$$

The reader will note that this is a fixed state feedback plus a constant (which depends upon the active constraint values). This feedback is feasible for all $\mathbf{x}$ such that the ignored constraints, that is $C_{in}\underset{\rightarrow test}{\Delta\mathbf{u}} - \mathbf{d}_{in}$, are feasible and optimal if in addition the associated lagrange multipliers are positive.

MPQP is therefore summarised (rather crudely) below:

Algorithm 12.1 Off-line:

1. *For all feasible active sets, define a region $S^{(i)}$ such that $\mathbf{x} \in S^{(i)}$ implies that the control trajectory $\underset{\rightarrow}{\Delta\mathbf{u}} = -K^{(i)}\mathbf{x} + \mathbf{k}^{(i)}$ is feasible and optimal.*

2. *Reduce the regions $S^{(i)}$ so that there is no overlap or duplication.*

[§]For instance, a typical choice makes use of $\log(\mathbf{d} - C\underset{\rightarrow}{\Delta\mathbf{u}})$.

On-line:

1. *Do set membership tests to locate in which set* $S^{(i)}$ *the current state lies.*

2. *Implement the associated control law (e.g. (12.6)).*

The advantage of this algorithm is that all possible control laws are defined off-line hence transferring on-line computation to off-line. This should enable a reduction in on-line computational demand as well as improving visibility (potential for off-line analysis) of the nonlinear constrained control law. The weakness of this algorithm is that there may be a very large number of possible active sets and this implies not only a large search to find the correct $S^{(i)}$ but also large data storage requirements. Current work is looking at how to improve efficiency; for instance, means of testing set membership by just a functional evaluation and hence also avoiding the need to store the region definitions.

> **Summary:** MPQP replaces the on-line QP optimisation by a series of set membership tests. These tests can be carried out efficiently but the potentially large number of alternative active sets is still an issue that needs to be fully resolved before this algorithm can be used for large problems.

12.1.4 Simple but suboptimal approaches

In the original constrained GPC paper [149] there is some discussion of how the optimum might be related to the constraints. It is shown that in some cases saturation control is optimal and in the worst case is still feasible [24]. Other papers (e.g. [130]) have also shown that, at times, one can find the global optimum (or a suboptimum but feasible trajectory) with just a few very simple tests (at most n_c iterations). However, there is not, to the authors knowledge, a neat summary in existence of how to form an optimal synergy between the QP problem and the MPC optimisation. It should be possible to use the special structure (such as constraint symmetry and Toeplitz structures in C) of the MPC problem to write an application specific QP algorithm that is highly efficient especially if one were to accept a small degree of suboptimality. For instance, one avenue being pursued within MPQP, to reduce the number of sets, is to say that only the first control move needs to be computed explicitly; one only needs to know that the remainder are feasible.

12.2 Introduction to computationally efficient MPC

The first section looked at the optimisation algorithm itself. The remainder of this chapter concentrates on how the MPC optimisation can be modified in order to bring about a reduction in the on-line computational load [2],[118],[146], [156].

12.2.1 Concepts used to reduced on-line computation

To reduce the computational burden associated to constraint handling one first needs a good understanding of what makes explicit constraint handling necessary. This leads to insight as to when explicit constraint handling can be avoided.

- Explicit constraint handling is required to avoid prediction mismatch (see Section 6.1) and hence to enhance performance.

- Explicit constraint handling is required to avoid infeasibility and hence potential instability/poor performance.

- Explicit constraint handling is not required when the state is inside the maximal admissible set (MAS) or other invariant set, so that unconstrained control satisfies constraints.

- Constraint satisfaction can at times be ensured by a simple set membership test rather than explicit comparison of predictions with constraints.

The aim now is to build on these observations in order to develop algorithms that reduce the on-line computational burden.

Summary:

- Constraint handling is not required inside the MAS (or any invariant set) associated to the applied control law.

- The key to eliminating on-line optimisation burden is to transfer as much as possible of the constraint handling to off-line computations.

12.2.2 Invariant sets

This section gives a brief review of invariant sets (see Chapter 11) which is required for developments hereafter.

12.2.2.1 Polyhedral invariant sets

If complex enough[¶] [33], polyhedral sets maximise the reachable space in the linear case. However, maximal volume is achieved at the cost of high complexity and these can be difficult to define in the presence of uncertainty.

[¶]Low dimensional polyhedral sets, hence conservative in the volumes contained, will not be considered here.

12.2.2.2 Ellipsoidal invariant sets

Ellipsoidal sets can be defined with only a few parameters, even in the uncertain case, but may be conservative in the volumes contained. They are not maximal in volume, even for symmetric constraints and can be particularly conservative for non-symmetric constraints. Hence not all initial states x_0 which if left to evolve freely given the state-feedback law $u_i = -K_0 x_i$ and model (2.16) and still satisfy constraints (12.8) are contained in the set. So, the use of ellipsoidal invariant sets may limit unnecessarily the available control authority, especially if the state feedback is fixed.

12.2.2.3 Can MPC be tuned systematically to change the volume of the associated invariant set?

For the simple case of open-loop stable systems which are subject to input constraints only, varying the control weighting in the MPC cost is sufficient to increase the volume of the MAS (e.g. [2], [118], [146], [156]. However, this does not apply to the general case with state constraints. The obvious question then is, how can one design the control law to maximise the volume of the associated MAS ? This is still an open question although it has been tackled for ellipsoidal invariant sets using LMI methods (e.g. [59, 123]).

> **Summary:** Constraint handling requirements are reduced if the terminal invariant set is as large as possible. However, one does not want to make the associated control law too detuned, as this detuning is partially inherited by the corresponding dual model MPC algorithm.

12.2.3 Methods illustrated in this chapter

This chapter will first demonstrate three approaches with negligible on-line optimisation. These methods are based on the off-line computation of large invariant sets and trade-off optimality for computational gains. A by-product of the reduction in computation and making more use of invariance is an increase in the transparency of how the controller works when constraints are active. One method [156], [146] will be called NESTED as it makes use of invariant sets which are preferably nested; the second method [129] is called ONEDOF as it deploys just one d.o.f. and the third is called efficient MPC [66] (EMPC). As this topic is becoming more specialised the algorithms will not be described in fine detail.

> **Summary:** The methods to be illustrated reduce computational load by making maximum use of the concept of invariance.

12.2.4 Notation and assumptions

For simplicity the ideas will be illustrated only on state-space models

$$\mathbf{x}_{k+1} = A\mathbf{x}_k + B\mathbf{u}_k; \quad \mathbf{y}_k = H\mathbf{x}_k \tag{12.7}$$

and perfect state knowledge is assumed. For possible extensions to deal with estimation errors see [20]. State and input constraints will be denoted by

$$C\mathbf{x} - \mathbf{d} \leq 0; \quad E\mathbf{u} - \mathbf{f} \leq 0 \tag{12.8}$$

Under the assumption of a fixed state feedback $\mathbf{u} = -K\mathbf{x}$, the two sets of constraints can be combined into:

$$G\mathbf{x} - \mathbf{h} \leq 0; \quad G = \begin{bmatrix} C \\ -EK \end{bmatrix}; \quad \mathbf{h} = \begin{bmatrix} \mathbf{d} \\ \mathbf{f} \end{bmatrix} \tag{12.9}$$

The underlying MPC strategy will be taken as the dual mode strategy of [137] (Section 7.3.4) implemented using the CLP. Hence for K_0 the optimal feedback:

$$\min_{\mathbf{c}_i, i=0,..,n_c-1} J_2 = \sum_{i=0}^{n_c-1} \mathbf{c}_i^T W \mathbf{c}_i \quad \text{s.t.} \quad \begin{cases} \mathbf{u}_i = -K_0\mathbf{x}_i + \mathbf{c}_i, & i = 0, 1, ..., n_c - 1 \\ \mathbf{u}_i = -K_0\mathbf{x}_i, & i \geq n_c \\ \text{constraints } (12.8) \\ \text{model } (12.7) \end{cases} \tag{12.10}$$

where W is defined from a corresponding infinite horizon cost; hence

$$W = R + B^T P_L B; \quad P_L = A^T P_L A - A^T P_L B (R + B^T P_L B)^{-1} B^T P_L A + Q \tag{12.11}$$

An ellipsoidal invariant set associated to K_0 will be defined as

$$S_E = \{\mathbf{x} : \mathbf{x}^T P \mathbf{x} \leq 1\}; \quad P > 0 \tag{12.12}$$

where $(A - BK_0)^T P(A - BK_0) \leq P$ and P small enough such that (12.8) is satisfied for all $\mathbf{x} \in \mathbf{S}$.

The MAS [33] S_P is defined as:

$$S_P = \{\mathbf{x} : G_a\mathbf{x} - \mathbf{h}_a \leq 0\}; \quad G_a = \begin{bmatrix} G \\ G\Phi \\ \vdots \\ G\Phi^{i-1} \end{bmatrix}; \quad \mathbf{h}_a = \begin{bmatrix} \mathbf{h} \\ \mathbf{h} \\ \vdots \\ \mathbf{h} \end{bmatrix} \tag{12.13}$$

Many of the constraints used in the definition of S_P may be redundant and should be removed off-line.

12.3 Three computationally efficient algorithms using invariant sets

The size of invariant sets varies enormously with the tuning of the control law. This very fact was fundamental in the algorithms proposed in [59, 118, 146, 156]. Three controllers based on this insight are illustrated next.

12.3.1 Algorithms of [146], [156] (NESTED)

Given the work of [2], it is logical to find K that optimises (12.10) for a range of R from the desired value R_0 to a very high value R_n. Let the controllers and invariant sets associated to choice R_i be K_i and S_i, respectively. It is reasonable (although not necessary) to expect that often $R_i < R_{i-1} \Rightarrow S_{i-1} \subset S_i$. Given that the desired choice for R is R_0, then all other choices lead to suboptimal controllers but in general allow the definition of larger invariant sets and hence extend the applicability of control. A logical control law [146] is therefore given as follows:

Algorithm 12.2 NESTED

1. *Let $i = 0$*

2. *Test whether $\mathbf{x} \in S_i$*

    (a) *If $\mathbf{x} \notin S_i$, then set $i = i + 1$ and go to step 2.*

    (b) *If $\mathbf{x} \in S_i$, let $K = K_i$.*

3. *Implement the control law $\mathbf{u} = -K\mathbf{x}$*

4. *Update sample instant and go to step 1.*

Remark 12.1 • *This control law is defined for all $\mathbf{x} \in S_0 \bigcup S_1 \bigcup \cdots \bigcup S_n$. For a stable plant, applicability can usually be widened by increasing R_n however the resulting controller K_n will be suboptimal.*

• *The algorithm may not deal well with open-loop unstable plant where a nonlinear control law is required to increase the stabilisable region.*

• *The same algorithm can also be developed using ellipsoidal invariant sets [153] and as such would more easily incorporate uncertainty.*

• *Performance can be considerably suboptimal although it will be reliable within the given sets.*

- *The main computational load is the set membership tests. Each of these could require significant numbers of multiplications and therefore one may not want a large number of sets or any computational benefit will be lost.*

Summary: NESTED makes use of a set of known control laws. Hence its behaviour can be analysed and made robust but it will be limited to the MAS of the given *linear* control laws.

12.3.2 One d.o.f. algorithms (ONEDOF)

One major shortcoming of NESTED is the potential conservatism. This is a consequence of not having any explicit optimisation in the on-line algorithm. If one were to allow some optimisation, albeit a trivial one, it is possible to reduce this conservatism considerably.

A simple algorithm ONEDOF using this philosophy was proposed in [64, 118, 129]. It is based on two control laws K_0, K_n, that is, the desired (highly tuned) control law and the law with maximal volume MAS. Despite the use of just a single degree of freedom and only two control laws, it frequently outperforms NESTED.

Algorithm 12.3 ONEDOF

1. *Define $\Phi_0 = A - BK_0$, $\Phi_n = A - BK_n$, denote by $\mathbf{x}_0$ the current state, and let the predicted state evolution with each control law be respectively,*

$$\mathbf{x}_k = \Phi_0^k \mathbf{x}_0; \quad \mathbf{x}_k = \Phi_n^k \mathbf{x}_0 \tag{12.14}$$

2. *Compute predicted inputs and corresponding state predictions (α a scalar)*

$$\mathbf{u}_k = -(1 - \alpha)K_0 \Phi_0^k \mathbf{x}_0 - \alpha K_n \Phi_n^k \mathbf{x}_0$$
$$\mathbf{x}_k = (1 - \alpha)\Phi_0^k \mathbf{x}_0 + \alpha \Phi_n^k \mathbf{x}_0 \tag{12.15}$$

3. *Substitute predictions (12.15) into the constraints $C\mathbf{x}_k - \mathbf{d} \leq 0$, $E\mathbf{u}_k - \mathbf{f} \leq 0$, $k = 0, 1, \dots$ and $0 \leq \alpha \leq 1$ to give feasible region*

$$\mathbf{m}\alpha - \mathbf{n} \leq 0 \tag{12.16}$$

where $\mathbf{m}$, $\mathbf{n}$ are linear in the initial state $\mathbf{x}_0$.

4. *Minimise α subject to (12.16). Use the optimum α to compute $\mathbf{u}_0$ from (12.15)*

5. *Update the sampling instant and go to step 2.*

Remark 12.2   • *The on-line optimisation is equivalent to minimisation$^\|$ of α w.r.t. the scalar α and subject to (12.16). This requires only a simple set of inequality checks and hence is trivial.*

$^\|$It is easy to show that minimising J with predictions (12.15) is equivalent to minimising α.

- *If* **x** *is well inside* S_n*, then the optimal* α *will be much less than* 1 *thereby indicating that and one can use a control law somewhat more highly tuned than* K_n *and still satisfy constraints.*

Hence α is used to gradually increase the tuning as **x** converges to the origin. In this sense it also has connections to [59] which designs a control law so that the state is always on the border of the invariant set for the control law selected.

> **Summary:** ONEDOF improves on NESTED by allowing less conservative behaviour when the state is not near a set boundary. Hence one can use far fewer sets. The implied computation is trivial and can be much less than NESTED.

12.3.3   Algorithm of [65]

The idea of using invariant sets of varying size allied to an implied control law was implicit in [59] but the proposed algorithm calculates the time varying control law on-line and hence requires a considerable amount of on-line computation in order to handle the performance minimisation simultaneously. It is possible to reduce this dramatically by transferring the burden from on-line to off-line computation [65, 66]. This approach is demonstrated next.

The main idea was to use the insight gained with the CLP and produce an invariant set which incorporated d.o.f. during the transients. Hence invariance is defined by the existence of a feasible control trajectory such that an *augmented* state remains within an invariant set. By allowing freedom within the transients of the control trajectory, as in dual mode control, the invariant set is enlarged and there is also more flexibility to optimise performance.

In order to present this algorithm the reader needs to be aware of how d.o.f. are used to form an augmented state and then how this is used to form an invariant set.

12.3.3.1   The autonomous model

Here we show how the ellipsoidal invariant sets (12.12) can be enlarged by adding d.o.f. with a fixed K. The d.o.f. c_i (see eqn.(12.10)) must be added to the state vector to create an autonomous system from the equations (12.7) and $u_k = -K x_k + c_k$, $c_k = 0, k \geq n_c$. This can be done as follows:

$$\mathbf{X}_{k+1} = \Psi \mathbf{X}_k; \quad \mathbf{X}_{k+1} = \begin{bmatrix} \mathbf{x}_{k+1} \\ \hline \mathbf{c}_1 \\ \mathbf{c}_2 \\ \vdots \\ \mathbf{c}_{n_c-1} \\ 0 \end{bmatrix}; \quad \mathbf{X}_k = \begin{bmatrix} \mathbf{x}_k \\ \hline \mathbf{c}_0 \\ \mathbf{c}_1 \\ \vdots \\ \mathbf{c}_{n_c-2} \\ \mathbf{c}_{n_c-1} \end{bmatrix}; \quad \Psi = \begin{bmatrix} \Phi & B & 0 \ldots 0 \\ \hline 0 & 0 & I \\ 0 & 0 & 0 \ldots 0 \end{bmatrix}$$

$$(12.17)$$

where $\Phi = A - BK$ and I is an identity matrix.

12.3.3.2 Setting up the invariant set

The d.o.f. $\mathbf{c}_i$ can be considered as states in an augmented autonomous model (12.17) and hence one can define an invariant set S_X for the expanded state $\mathbf{X}$; this is chosen so that $\mathbf{X} \in S_X$ implies predicted constraint satisfaction and convergence (following similar lines to those in Section 11.9.2).

S_X has an obvious projection S_E onto $\mathbf{x}$-space and thus for all $\mathbf{x} \in S_E$, there must exist a set of d.o.f. $\mathbf{C} = [\mathbf{c}_0^T, ..., \mathbf{c}_{n_c-1}^T]^T$ such that $\mathbf{X} \in S_X$, i.e. giving feasible predictions. What remains therefore is to show an efficient means of selecting the d.o.f. $\mathbf{C}$ so as to optimise J of (12.10).

Remark 12.3 *The volume of S_E increases with n_c; that is, one can use the d.o.f. to enlarge the feasible region. However, the complexity of description of S_E cannot change so the only cost of increasing the number of d.o.f. n_c is the off-line computational load.*

12.3.3.3 EMPC algorithm

Algorithm 12.4 EMPCold

1. *Define invariant set $S_X = \{\mathbf{X} : \mathbf{X}^T P_X \mathbf{X} \leq 1\}$ and its projection to x-space $S_E = \{\mathbf{x} : \mathbf{x}^T P \mathbf{x} \leq 1\}$ for given n_c and also the invariant set S_{E0} for $n_c = 0$.*

2. *If the current state $\mathbf{x}_0 \in S_{E0}$, $\mathbf{u} = -K_0 \mathbf{x}$ is feasible. Use $\mathbf{u}_0 = -K_0 \mathbf{x}_0$.*

3. *If $\mathbf{x}_0 \notin S_{E0}$,*

    (a) Minimise w.r.t. $\mathbf{C}$ the J of (12.10) s.t. $\mathbf{X} \in S_X$.

    (b) Of the optimum $\mathbf{C}$, use $\mathbf{c}_0$ to compute current control $\mathbf{u}_0 = -K_0 \mathbf{x}_0 + \mathbf{c}_0$.

4. *Update sample instant and go to step 2.*

Remark 12.4 *The computation implied in step 3a above is trivial [66]: it reduces to finding the only positive real root of a polynomial. Moreover, the online objective is to minimise the desired J; that is, unlike with NESTED, one maintains an explicit handle on performance.*

Remark 12.5 *Handling constraints through the use of ellipsoidal sets is conservative. This can be reduced by scaling $\mathbf{C}$ thereby allowing $\mathbf{X}_0$ to move outside S_X; the lost guarantee of feasibility can be regained by requiring $\mathbf{X}_1 \in S_X$ and ensuring that constraints are satisfied at current time. This requires the solution of the roots of a quadratic [65].*

Summary: EMPC allows systematic incorporation of the d.o.f. during transients which increases the feasible region, but it is restricted to ellipsoidal regions. The implied on-line computation is the solution of the only positive real root of a polynomial which is trivial.

12.3.4   Overview of QPMPC, NESTED, ONEDOF, EMPC

- The transparency of NESTED enhances the reliablity of implementation, but:

  1. It has restricted applicability and may not extend easily to cases with state with constraints.

  2. The control can be significantly suboptimal unless many sets are used which could make it computationally demanding.

- ONEDOF has similar feasibility regions to NESTED but:

  1. It allows a systematic improvement in performance and a reduction in the computational load.

  2. The analysis of performance is less straightforward.

- EMPC performs an explicit, but trivial, optimization of the true cost J and hence is the most optimal.

  1. It allows systematic inclusion of d.o.f. in transient which increases feasible regions.

  2. It is restricted to ellipsoidal sets which limits applicability.

  3. The optimisation, although simple, is more demanding than ONEDOF.

It remains now to illustrate these three algorithms on some numerical examples, and in particular, to investigate how much suboptimality has been traded in return for the reduction in on-line computation.

Summary: The main price of computational simplicity is a reduction in the size of the feasible region. This is because there are less d.o.f. to give freedom in the predictions. There will also be some suboptimality due to the implied restriction in the parameterisation of the future control trajectory.

12.3.5   Examples

The purpose of this section is to illustrate, by simulation, the algorithms of NESTED, EMPC and ONEDOF and contrast their performance to the global optimum (LQMPC of Section 7.3.4 with large n_c). Examples are restricted to second order systems for which it is possible to plot feasible regions.

For the NESTED algorithm, the maximal volume set will correspond to the optimal control gains obtained for control weighting $R * 100$ where R will be taken to be $R = 0.1$.

12.3.5.1 Example 1 – A double integrator

$$A = \begin{bmatrix} 1 & .1 \\ 0 & 1 \end{bmatrix}; \quad B = \begin{bmatrix} 0 \\ 0.0787 \end{bmatrix}; \quad H = \begin{bmatrix} 1 & 0 \end{bmatrix} \qquad (12.18)$$

with constraint matrices

$$C = \begin{bmatrix} 1 & -0.333 \\ -0.333 & 1 \\ -1 & 0.333 \\ 0.333 & -1 \end{bmatrix}; \quad d = \begin{bmatrix} 1 \\ 1 \\ 1 \\ 1 \end{bmatrix}; \quad E = \begin{bmatrix} 1 \\ -1 \end{bmatrix}; \quad f = \begin{bmatrix} 1 \\ 1 \end{bmatrix} \qquad (12.19)$$

Using $Q = C^T C$ and $R = 0.1$ and $n_c = 3$.

12.3.5.2 Example 2 – Unstable process

$$A = \begin{bmatrix} 1.2 & .1 \\ 0 & 1 \end{bmatrix}; \quad B = \begin{bmatrix} 0 \\ 0.0787 \end{bmatrix}; \quad H = \begin{bmatrix} 0.1 & 2 \end{bmatrix} \qquad (12.20)$$

with constraint matrices

$$C = \begin{bmatrix} 1 & -0.333 \\ -0.333 & 1 \\ -1 & 0.333 \\ 0.333 & -1 \end{bmatrix}; \quad d = \begin{bmatrix} 1 \\ 1 \\ 1 \\ 1 \end{bmatrix}; \quad E = \begin{bmatrix} 1 \\ -1 \end{bmatrix}; \quad f = \begin{bmatrix} 1 \\ 1 \end{bmatrix} \qquad (12.21)$$

Using $Q = C^T C$ and $R = 0.1$ and $n_c = 3$.

12.3.5.3 Regions of attraction

The regions of attraction for the algorithms are plotted in Figures 12.1, 12.2 and 12.3 for examples 1 and 2, respectively. The regions for NESTED (which are the same for ONEDOF) are in dotted lines; for EMPC are in dashed lines; and for QPMPC in dash-dot lines. For completeness the maximal admissible set for $n_c = 0$ is shaded.

For example 1 (Figure 12.1) it is clear that increasing the control weighting alone is a quite effective way of enlarging the region of attraction and in fact this is far more effective than EMPC.

For example 2 Figure 12.2 (only half the regions are plotted for clarity – the plot is symmetrical), a very different picture emerges. In this instance, NESTED has performed poorly and in fact although increasing the control weighting enlarges the overall region of attraction, in some directions it has contracted. EMPC gives a slightly fatter set.

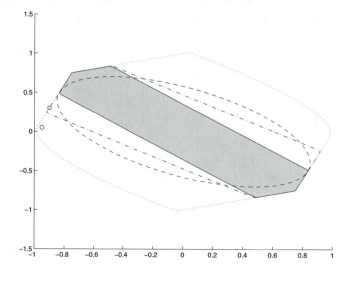

FIGURE 12.1

Regions of attraction for model (12.18).

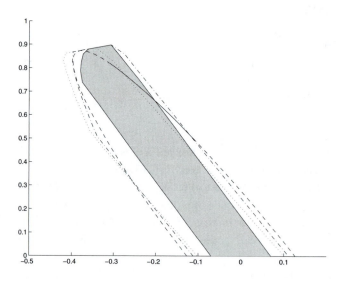

FIGURE 12.2

Regions of attraction for model (12.20) with state constraints.

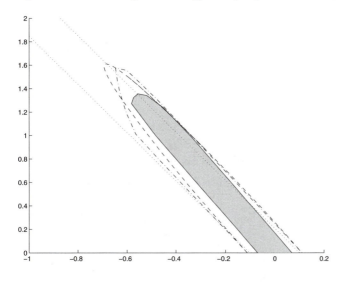

FIGURE 12.3
Regions of attraction for model (12.20) without state constraints.

Summary: Although feasible regions can be increased by changes in R or using the autonomous model formulation, the improvements can be poor when compared to the more usual LQMPC algorithm (dash-dot lines).

12.3.5.4 Closed-loop performance - simulations

To make the comparison more meaningful it is necessary to consider initial conditions which lie inside the regions of attraction of all the algorithms under consideration.

Here the plots are given for just one example, as these are illustrative of all the others. Take example 1 and an initial state (marked with a square in Figure 12.1) Base the NESTED algorithm on 3 sets for $R = 0.1, 1, 10$. The input, output and state trajectories are displayed in Figures 12.4, 12.5 where the solid lines are used for LQMPC, the dotted lines are used for NESTED and the dashed lines are used to denote constraints. The simulations for LQMPC, ONEDOF and EMPC are indistinguishable but there is clearly a sudden change in the NESTED input plot; this corresponds to a switching from control law K_1 to K_0 and is evidence of the suboptimality in the algorithm.

Summary: NESTED shows a distinct change in control (Figure 12.4) and therefore suboptimality. EMPC and ONEDOF are often able to come arbitrarily closed to the global optimal with a negligible computational burden.

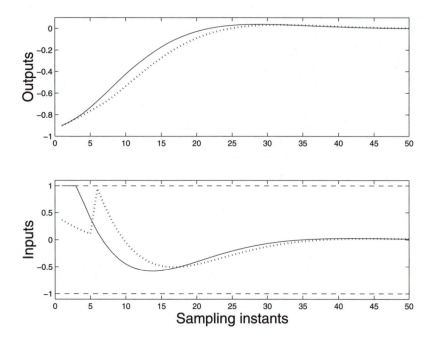

FIGURE 12.4

Output and input trajectories for example 1.

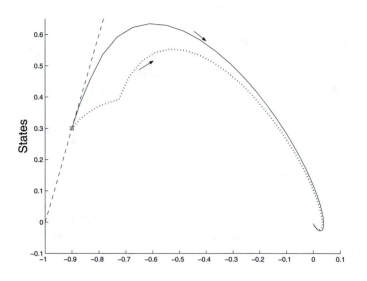

FIGURE 12.5

State trajectories for example 1.

TABLE 12.1

Run time costs

x_0	LQMPC	Optimal	ONEDOF	NESTED	EMPC
$[-0.9,0.3]^T$	6.112	6.112	6.112	6.867	6.112
$[-0.95,0.05]^T$	infeas	9.464	9.464	10.896	infeas
$[-0.36,0.55]^T$	infeas	2.9434	2.9771	3.419	infeas
	(2.9415 for $n_c \geq 8$)				

12.3.5.5   Run time costs

The run time costs are computed for some closed-loop simulations $x_0 = [-0.9,0.3]^T$, $[-0.95,0.05]^T$, $[-0.36,0.55]^T$ respectively and tabulated in Table 12.1. Where marked as such, LQMPC is infeasible, for $n_c = 3$; the global optimum assumes large n_c.

- In the second test, LQMPC and EMPC are infeasible. Nevertheless ONEDOF is indistinguishable from the global optimum! NESTED is again suboptimal.

- If NESTED makes use of more sets, e.g. using $R = 0.1,\ 0.3,\ 1,\ 3,\ 10$ the associated run time cost is to 6.246, 10.161 for simulations 1, 2 respectively; that is, more sets imply better optimality.

- In the third test, LQMPC (with 3 d.o.f.) and EMPC are infeasible. ONEDOF has performed very well and has near optimal responses whereas NESTED is nearly 20% worse. Performance with NESTED can be improved by increasing the number of sets.

> **Summary:**
>
> 1. Where it is feasible, ONEDOF often gives performance close to the global optimum. This is particular useful in cases where retuning R is a more effective way of increasing feasible regions than increasing n_c.
>
> 2. EMPC is effective where feasible but the restriction to ellipsoidal sets is a problem.

12.3.5.6   Computational load comparisons

NESTED: Has a low load if a small number of sets are deployed but this results in poor performance. Good performance requires more sets and hence a high computational load.

ONEDOF: has a very low computational load – essentially the update of vectors m, n in (12.16).

EMPC: Has a larger load than ONEDOF but with no increase in complexity can be applied to the robust case!

12.3.6 Summary

This section has illustrated some possible means of reducing computational load within MPC while still doing constraint handling. Unsurprisingly there is a price to pay either in reduced feasibility regions and/or suboptimality. There are few results in the literature looking at this area, as it is assumed that for most processes the computational demand of a QP is not a problem. This may change if MPC is to become more widely used, such as on processes with sample times of milliseconds.

> **Summary:** It has been shown that algorithms such as NESTED and, in particular, ONEDOF can give near optimal control with a low load where retuning R is an effective means of forming large invariant sets. However, there remains the fundamental problem of how to form large invariant sets, when retuning R is ineffective.

12.4 Stability analysis and options within ONEDOF

The previous section did not discuss stability. In is easy to show that NESTED and EMPC** are guaranteed stabilising for the nominal case. However, this is not the case for ONEDOF and hence some brief discussion is given in this section for completeness. Second, ONEDOF was presented as a modification of NESTED, however other equally effective alternatives exist and so again for completeness some discussion is given here.

12.4.1 Stability of ONEDOF using LMIs

The easiest stability analysis for ONEDOF is based on invariant sets and is a straightforward application of ideas used in [5, 59] and also given in Section 11.10.2.2. Consider the two control laws (12.15) within ONEDOF. These laws, when combined with model (12.7) and used alone, give rise to two possible closed-loop systems:

$$
\begin{aligned}
\mathbf{x}_{k+1} &= \Phi \mathbf{x}_k; & \Phi &= A - BK \\
\mathbf{x}_{k+1} &= \Phi_n \mathbf{x}_k; & \Phi_n &= A - BK_n
\end{aligned}
\tag{12.22}
$$

**In fact EMPC can relatively easily be applied to the uncertain case too as can NESTED with suitable set definitions.

Now, the current control is given from

$$\mathbf{u} = [-(1-\alpha)K + \alpha K_n]\mathbf{x} \qquad (12.23)$$

where α is unknown a priori. Therefore the true system closed-loop could be arbitrarily time varying within the convex hull of Φ, Φ_n:

$$\mathbf{x}_{k+1} = \Psi\mathbf{x}_k; \quad \Psi = (1-\alpha_k)\Phi + \alpha_k\Phi_n, \ \ 0 \le \alpha_k \le 1 \qquad (12.24)$$

A sufficient (though not necessary) test for the convergence of $\mathbf{x}_k$ in such a system is given by the following LMI invariance conditions:

$$\begin{array}{c} \Phi^T P \Phi - P \le 0 \\ \tilde{\Phi}^T P \tilde{\Phi} - P \le 0 \end{array}; \quad P > 0 \qquad (12.25)$$

The existence of any P satisfying (12.25) would therefore verify that the system (12.24) was stable for any time variation of α.

Remark 12.6 *If no solution exists to (12.25), this does not mean that the control law will be detabilising; rather it only shows that we have not yet proved it is stabilising. If a proof is desirable, then the optimal unconstrained control law K would need to be detuned towards K_n until a suitable invariant set can be defined. Less conservative tests for stability are possible [128] but too advanced for this book.*

Summary: Simple LMI tests can be use to analyse the stability of ONEDOF; these, however, are sufficient but not necessary. This author has yet to find examples where ONEDOF did not perform well.

12.4.2 Options in ONEDOF

Philosophically the key component in ONEDOF can be viewed as the definition of a better parameterisation of the d.o.f. for control; that is, interpolate between whole trajectories which have good properties.

The main idea used in the early papers [64, 118] was to define and use, in an efficient manner, two control trajectories:

- The first satisfied constraints and asymptotically the model output trajectories converged. Define such a future control trajectory as $\mathbf{u}_{feas}$.

$$\mathbf{u}_{feas} = \tilde{\mathbf{u}}_{k|k}, \tilde{\mathbf{u}}_{k+1|k}, \cdots \qquad (12.26)$$

- The second gives optimal performance in the absence of constraints:

$$\mathbf{u}_{opt} = \hat{\mathbf{u}}_{k|k}, \hat{\mathbf{u}}_{k+1|k}, \cdots \qquad (12.27)$$

The desire is to use $\mathbf{u}_{opt}$ where this is feasible. If it is not, then $\mathbf{u}_{opt}$ and $\mathbf{u}_{feas}$ are combined so as to optimise predicted performance s.t. constraints being satisfied. Hence define the predicted input trajectory as:

$$\mathbf{u}_{mix} = (1 - \alpha_k)\mathbf{u}_{opt} + \alpha_k\mathbf{u}_{feas}; \quad 0 \le \alpha_k \le 1 \qquad (12.28)$$

where clearly (12.28) is guaranteed to contain at least one feasible choice. Substituting $\mathbf{u}_{mix}$ into (12.10, 12.7) it is easy to show that J is minimised by minimising α_k s.t. constraints, that is, a linear program (LP) in just one variable.

What remains is to ask how $\mathbf{u}_{feas}$ might be determined, as this is the main component of ONEDOF. Some different alternatives are outlined next.

Summary: ONEDOF requires a feasible trajectory. The feasible region and computational load of the algorithm depend upon how this feasible trajectory is defined.

---

1. *The tail*: The tail was discussed in Section 6.3.1. It was shown that including the tail [64] as a possible choice of the new predicted control trajectory facilitated a convenient proof of stability because this enabled a guarantee both of feasibility and a reduction in the cost. The weakness of this choice is that the feasibility of the tail is not robust to even small uncertainty, as the tail could be on a constraint boundary and takes no account of new measurements.

2. *Mean level:* The mean level choice [118] is similar to the sort of philosophy adopted in [16, 59]. That is, as the control law is made successively more cautious (equivalent to increasing R in (12.10)), then the control moves become smaller and hence the unconstrained control law is more likely to be feasible and satisfy (12.8). This choice is also robust to some parameter/signal uncertainty. The weakness of this choice is that the extension to cater for state constraints is not obvious.

3. *Open-loop response:* This choice is crude but simple [121, 131] and it is restricted to the case of stable open-loop plant. It reduces to low gain integral control, that is choose

$$\mathbf{u}_{feas} = [\breve{\mathbf{u}}, \breve{\mathbf{u}}, \ldots]; \quad \breve{\mathbf{u}} = \frac{\mathbf{u}_\infty}{\beta} + \frac{(\beta - 1)\mathbf{u}_{k-1}}{\beta}; \quad \beta \ge 1 \qquad (12.29)$$

This means that the input gradually creeps toward the predicted steady-state value without any overshoots. This can be represented as a state feedback but the weakness, as with mean level, is that it does not easily take account of state constraints.

4. *Reduced order MPQP [134]:* This choice constitutes work in progress. The idea is to find the maximal volume feasible region for a given n_c and then parameterise the boundary of this region. This should entail far fewer regions than MPQP. One would then associate the current state measurement to a

boundary and hence identify quickly an associated feasible trajectory. The advantage is that the feasible region is large but one may inherit a large overhead in data storage and searching over sets, as with the full MPQP.

Summary: Many options exist within ONEDOF. If the process can only cope with a limited computational burden, then it is worth investigating whether ONEDOF has enough flexibility to handle the problem. The implied coding and on-line computation are very small compared to conventional MPC allied to a QP.

12.5 Conclusions

It is possible to formulate MPC algorithms with a relatively low computational burden. The key concepts implicit in this are:

- Transference of on-line optimisation to off-line computations and analysis is used.

- Membership of an invariant set is used to test for constraint satisfaction.

- Alternative parameterisations of the d.o.f. in the optimisation. Use whole trajectories rather than individual values.

  1. In NESTED the d.o.f. are the choice of underlying control law.

  2. In ONEDOF the d.o.f. is an interpolation between two possible input trajectories.

  3. In EMPC the d.o.f. are absorbed into the model and the optimisation is changed from a QP to a simpler one.

- Suboptimality is implicit due to the restriction in the parameterisation of the d.o.f. and hence also the implied restriction in the volume of feasible region.

Summary: One can often get very good performance and constraint handling with recursive feasibility while using a relatively simple MPC algorithm.

13

Predictive functional control

Predictive functional control (PFC) is the product of a company, ADERSA, which has a main aim of maximising the takeup of MPC within industry. It achieves this partially by keeping the algorithm as simple as possible and hence this is a logical algorithm to follow the previous chapter.

As with other predictive controllers, PFC is based on the use of predictions and hence a prediction model; however, here the similarities stop. The design method and the constraint handling are much less sophisticated than with typical MPC algorithms and as a consequence the algorithm is far simpler. This simplification reduces the on-line computational demand significantly, hence opening the door to fast applications and nonlienar processes. Moreover the design is based on variables which engineers can relate to easily such as the desired rise time – as such tuning can be easier.

Of course the price one pays for this simplification is a loss of rigor in that one cannot give a priori guarantees of performance, especially during constraint handling. Nevertheless the hundreds of successful implementations are sufficient evidence that the strategy is often effective in practice. It should be emphasised that PFC is really only suitable for SISO loops.

Summary: This chapter gives a brief overview of the key components in PFC and illustrations of the efficacy of the approach despite its apparent simplicity.

13.1   Summary of the overall philosophy of PFC

This section introduces the key building blocks behind PFC. This will emulate to some extent the building blocks of any MPC algorithm, that is, predictions, objective, parameterisation of the d.o.f. and a receding horizon. So here the differences from more typical MPC algorithms will be emphasised.

The aim in PFC is to get the system response to emulate that of a first order system (with the suitable delay). Hence the design objective is parameterised by just two variables: the delay which is given as that of the process and the desired time constant for the closed-loop response. As such the design process is very easy for practising engineers who have a good feel for what is a realistic closed-loop bandwidth.

The design is computed by minimising the difference between the model predictions and that of the selected first order target. This is clearly the same type of aim that is used in conventional MPC algorithms*. However, typically in for instance GPC, this cost function leads to an optimisation (say least squares) which hence entails a significant computational burden. PFC avoids this burden by only comparing errors at the same number of points as there are d.o.f. in the input trajectory. The implication therefore is that one can make the predicted errors zero at the selected points (denoted coincidence points). Hence the optimisation reduces to the solution of a number of equality conditions which is equivalent to solving linear simultaneous equations; that is, there is no optimisation and therefore only trivial computations are required.

As with GPC/DMC, the nominal control law can lead to constraint violations and a simple deployment of saturation control may not be a good choice. Hence PFC has a slightly more advanced mechanism for constraint handling, although still suboptimal. In fact the method used is very similar to that of NESTED (Section 12.3.1). If the *best* PFC control law is expected to give constraint violations, then switch to an alternative law which is expected to be feasible. Of course this implies that an alternative strategy is known; in general the implied alternative trajectories would have to contain none overshooting behaviour.

Summary: The focus in PFC is simplicity and intuition. This facilitates acceptance by practising engineers and implementation on fast systems.

13.2   Derivation of the PFC control law

This section gives the mathematical details required to derive a PFC control law. Hence it gives details of the prediction model, the objective, the parameterisation of the d.o.f. and the *optimisation*.

13.2.1   Modelling and prediction

There is a preference in ADERSA for the use of the independent model (IM – see [30, 98] and Section 3.6). However, as discussed in chapter 10, the choice of prediction model has no effect on nominal tracking and rather affects loop sensitivity. Hence such issues are not considered in this chapter except where of significance.

---

*Original papers such as [21] allowed the incorporation of such filtered objectives but these gradually dropped out of the popular academic literature.

The predictions can be obtained in identical fashion to that detailed in Chapter 3. For instance for an input/output model the predictions could be given as (3.21) and for an IM the predictions could be given from any of (3.63, 3.64, 3.75) as relevant.

13.2.2 Desired reference trajectory

PFC places the desired closed-loop dynamic into the reference trajectory. Given the actual set point is r, the loop set point w is a first order lag and can be[†] defined as follows:

$$w_{k+i|k} = r_k - (r_k - y_k)\Psi^i \tag{13.1}$$

whre y_k is the most recent measured output and Ψ is a tuning parameter setting the desired closed-loop pole (this is related to the implied time constant T_d for the lag through $\Psi = e^{-T/T_d}$, T the sample rate).

The pseudo set point $w_{k+i|k}$, $i > 0$ follows an exponential curve from the current output to the desired asymptotic value of the output. One chooses the curvature, that is Ψ, according to plant requirements.

If there is a delay in the system, one could build this into the reference trajectory or use an undelayed output from the IM.

13.2.3 The coincidence points

The control law is determined by using the d.o.f. to enforce equality of the predictions and the reference trajectory at a number of points, that is, by solving for the future control moves such that:

$$y_{k+n} = w_{k+n}, \quad n = n_1, n_2, \ldots \tag{13.2}$$

These equalities are called coincidence points (as the predicted output and set point coincide). Typically there are only one or two coincidence points.

One will note that no criteria are placed on the input activity. It is assumed that the output tracking performance is the dominant criterion and the input should be selected as required to meet this. One can, however, limit input action by a wise parameterisation of the allowed predicted control trajectories.

13.2.4 Parameterisation of the d.o.f./future control trajectory

One significant difference between PFC and conventional MPC is the parameterisation of the predicted input trajectory. Rather than using the control moves themselves, PFC takes the trajectory as the sum of a step change, a ramp, a parabola, etc.

[†]Omitting details of the delay for simplicity.

The precise components to be included are selected to match the expected characteristics in the set point.

- If only step changes are expected, the input is parameterised as only the current
  increment in control.

- If ramp changes are expected, the input is parameterised as the current increment in control and a ramp rate on the input.

- If parabolic changes are expected, the input is parameterised as the current
  increment in control, a ramp rate on the input and a parabolic component.

The advantage of this form of parameterisation is that one can achieve no lag in the
response (asymptotically). An illustration is given next.

The predictions (e.g. (3.21)) have a component $H\Delta\underrightarrow{u}$ which details the impact of
future control moves. For a step and a ramp rate one would get

$$\Delta\underrightarrow{u} = \begin{bmatrix} \Delta u_k \\ 0 \\ 0 \\ \vdots \end{bmatrix} + \begin{bmatrix} 1 \\ 1 \\ 1 \\ \vdots \end{bmatrix}\beta \tag{13.3}$$

where β is the ramp rate. Hence the component of the prediction equations (3.21)
containing the d.o.f. would be altered as follows:

$$H\Delta\underrightarrow{u} = H\left(\begin{bmatrix} \Delta u_k \\ 0 \\ 0 \\ \vdots \end{bmatrix} + \begin{bmatrix} 1 \\ 1 \\ 1 \\ \vdots \end{bmatrix}\beta\right) = [H_1, HE]\begin{bmatrix} \Delta u \\ \beta \end{bmatrix} \tag{13.4}$$

where H_1 is the first column of H and E is a column vector of ones.

13.2.5   The control law

The control law is determined by solving the identities (13.2). Define the n_1^{th} standard
basis vector as $\mathbf{e}_{n_1}$; then (using prediction model (3.21)) one can compute the n_1^{th} step
ahead prediction

$$y_{k+n_1} = \mathbf{e}_{n_1}^T \underrightarrow{y} = \mathbf{e}_{n_1}^T[H\Delta\underrightarrow{u} + P\Delta\underleftarrow{u} + Q\underleftarrow{y}] \tag{13.5}$$

Algorithm 13.1 *PFC with one coincidence point can be determined by substituting
prediction (13.5) into coincidence condition (13.2) and then substituting in reference
trajectory (13.1) and (13.4). Hence*

$$\mathbf{e}_{n_1}^T[H_1\Delta u + P\Delta\underleftarrow{u} + Q\underleftarrow{y}] = r_k - (r_k - y_k)\Psi^{n_1} \tag{13.6}$$

which implies that the control law is given from

$$\mathbf{e}_{n_1}^T H_1 \Delta u = r_k - (r_k - y_k)\Psi^{n_1} - \mathbf{e}_{n_1}^T [P\Delta\underleftarrow{u} + Q\underleftarrow{y}] \tag{13.7}$$

As in Section 4.3, this can easily be expressed as a fixed linear feedback law in the form of (4.25). Hence conventional a posteriori stability and sensitivity analysis (e.g. Chapter 9) can be applied in a straightforward manner.

Remark 13.1 *For two coincidence points (using (13.4)) the control law is given from*

$$\mathbf{e}_{n_1}^T \{[H_1, HE] \begin{bmatrix} \Delta u \\ \beta \end{bmatrix} + P\Delta\underleftarrow{u} + Q\underleftarrow{y}\} = r_k - (r_k - y_k)\Psi^i \tag{13.8}$$

This implies that

$$\mathbf{e}_{n_1}^T [H_1, HE] \begin{bmatrix} \Delta u \\ \beta \end{bmatrix} = \begin{bmatrix} r_k - (r_k - y_k)\Psi^{n_1} \\ r_k - (r_k - y_k)\Psi^{n_2} \end{bmatrix} - \begin{bmatrix} \mathbf{e}_{n_1}^T \\ \mathbf{e}_{n_2}^T \end{bmatrix} [P\Delta\underleftarrow{u} + Q\underleftarrow{y}] \tag{13.9}$$

and the first control move is $\Delta u + \beta$.

Summary: PFC replaces objective optimisation by forcing a subset of the predictions to exactly match a modified set point trajectory. The d.o.f. are parameterised in an equivalent way to the set point. The on-line computational burden is negligible and gives rise a fixed linear control law.

13.3 Tuning PFC

The tuning parameters are the coincidence horizons, e.g. $n_1, n_2, \ldots$, and the desired time constant (or equivalently Ψ). In general it is difficult to see an intuitive argument for the tuning of the coincidence horizon and it is better to see this as a tool for achieving the specified closed-loop dynamics. Hence a typical procedure with one coincidence point would be as follows:

1. Choose the desired Ψ.

2. Do a search for $n_1 = 1, \ldots$, large and find the associated control law (13.7) for each n_1.

3. Select the n_1 which gives closed-loop dynamics closest[‡] to the chosen Ψ.

[‡]Use the arguments of Section 6.1 on the desire to minimise the mismatch between predictions and expected closed-loop behaviour.

4. Simulate the proposed law. If satisfactory exit; otherwise reselect Ψ and go to step 2.

Hence the tuning reduces to a global search, but this requires only relatively trivial computations and hence would be quite quick. With two coincidence points, the global search would be more involved but should still be quick.

A few heuristics would be needed to choose the *optimum* Ψ but the user can otherwise be distanced from the search for the best coincidence points. That is, the user chooses the desired Ψ and then is given the corresponding *best* PFC controller which can either be accepted or rejected.

Remark 13.2 *Some insight is possible into the impact of changing the coincidence points. For stable open-loop processes with one coincidence point:*

1. *If n_1 is large, then the control law will reduce to open-loop dynamics with integral action (because one will be choosing a single control move to eliminate the expected steady-state error). It will hence be guaranteed stable.*

2. *If n_1 is small, PFC is well posed only if closed-loop behaviour close to a first order lag is realistic. This follows from arguments of prediction mismatch (Section 6.1).*

3. *From point 1 it can be seen that Ψ is an effective tuning parameter only if n_1 is small. For n_1 large the resulting control law will be affected only a little by changes in Ψ. By n_1 small we mean that $\Psi^{n_1} > 0.2$.*

Summary: The tuning reduces to choosing the desired loop pole and then a global search to discern whether a PFC control law will give dynamics close to the requested ones.

13.4   Constraint handling

PFC has a limited systematic constraint handling facility as, unlike in GPC, there is no flexibility in the d.o.f. in the solution of the equality conditions (13.2). Hence one has no obvious mechanism for changing the selected $\Delta \underrightarrow{u}$ should this be infeasible. The only option therefore is to change the control law; that is, the underlying linear control law is changed, on-line, to cater for different scenarios.

This thinking is analogous to the NESTED algorithm of Section 12.3.1 although PFC would tend to use only two laws, much like algorithm ONEDOF. The resulting algorithm can be constructed as follows:

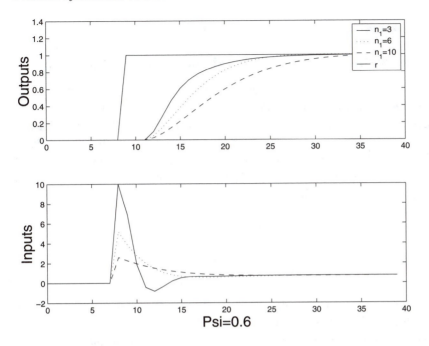

FIGURE 13.1

Closed-loop step responses for $\Psi = 0.6$.

1. Find the MAS for the desired control law and a detuned alternative.

2. If the state is not inside the MAS for the 'optimum' law [§], then use the alternative control law.

The weakness of this method, as for NESTED and ONEDOF, is that there is no obvious mechansim for finding the alternative detuned control law. The one obvious exception is stable processes without tight state constraints; in this case it is easy to prove that low gain integral control is always feasible.

> **Summary:** Effective constraint handling relies on the ability to fall back to a fixed linear control law that is known never to be infeasible. For complex dynamics this law may be hard to determine.

[§]That is constraint violations are predicted.

13.5   Simulation examples with a single coincidence point

In this section it will be illustrated just how effective PFC can be despite all the apparent theoretical shortcomings indicated above. As the constrained case has been discussed at depth in Chapter 12, this section will restrict itself to constraint free simulations. The aim is to show that with the relatively simple objective (a single coincidence point), performance may still be close to that achievable with an optimal control law such as LQMPC. For brevity just one example is shown; it is obvious that counter examples will exist so it is required only to show the PFC can be effective on some examples.

Take an example with a dead time of four samples:

$$G(z) = \frac{z^{-4}[0.1 + 0.003z^{-1}]}{1 - 1.8z^{-1} + 0.81z^{-2}} \tag{13.10}$$

Plot the closed-loop responses for $n_1 = 3$, 6, 10 with $\Psi = 0.6$, 0.8, 0.95. These simulations are given in Figures 13.1–13.3.

Changing Ψ: It is clear that as Ψ is increased, the corresponding closed-loop repsonses get slower. Hence Ψ can be used as a systematic tuning parameter for closed-loop behaviour.

Changing n_1: As n_1 is increased the behaviour tends towards open-loop (independent of Ψ). Note that for $n_1 = 10$, the input becomes almost a simple step to the required steady-state value. In this case lower values of n_1 give more aggressive control.

Summary: The PFC tuning parameters lend themselves to simple intuition and hence may be popular with practising engineers.

13.6   Unstable open-loop problems

Unstable processes can be difficult to control [143] and yet are quite common in some industries. For instance, exothermic batch reactors which also have a non-minimum phase characteristic due to the implied cooling when reactants are added or flexible beams which have oscillatory behaviour. There is a need for systematic control design tools to handle *complex* instability, that is, where PID design is non-trivial or fails[¶]. Moreover, for facilitating application to fast systems, these tools should do

---

[¶]One such example is a process with a factor of the form $(s - a)/(s - ra)$, $r > 1$.

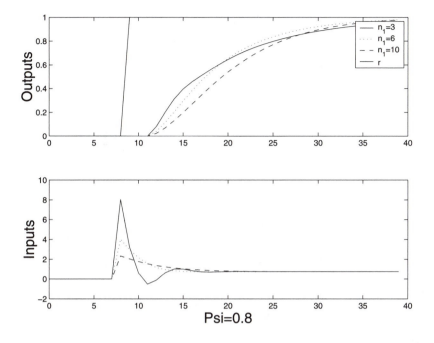

FIGURE 13.2

Closed-loop step responses for $\Psi = 0.8$.

systematic constraint handling but without requiring a high on-line computational load.

PFC would seem an obvious tool to try and in fact ADERSA have successfully applied PFC on many unstable systems. However, on others PFC performs poorly and hence there is a need to understand clearly the limitations to its use.

13.6.1 Weaknesses of PFC when applied to unstable processes

The weakness of PFC is exposed most clearly with either nonminimum phase processes and some unstable processes. Unsurprisingly, it is the identical problem that was identified for conventional MPC algorithms in Section 6.1, that is prediction mismatch. In summary, PFC will lead to good closed-loop performance if the predictions used are a good match to the consequent closed-loop behaviour; that is, if forcing coincidence in the predictions is nearly equivalent to obtaining coincidence in the closed-loop behaviour. Here, however, is a major problem.

- For nonminimum phase systems, the coincidence horizon would have to be large enough to occur beyond the nonminimum phase characteristic.

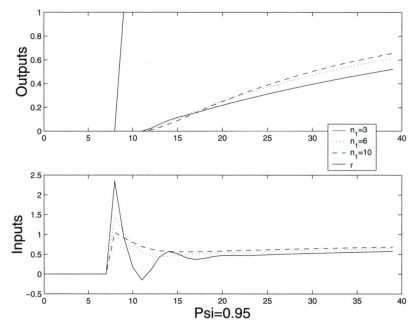

FIGURE 13.3
Closed-loop step responses for $\Psi = 0.95$.

- For unstable systems, especially with a single coincidence point, the predictions must be unstable and yet the desired behaviour is stable; that is, there is significant mismatch.

This section gives brief overview of how one can modify PFC so that it can handle unstable systems more reliably. The motivations for such extensions are:

1. PFC is popular in industry; hence any developments to extend its capacity are useful.

2. PFC is intuitive and computationally simple and hence has many applications on fast systems. This is an area yet to be well exploited by more conventional MPC algorithms.

Summary: A standard PFC algorithm can fail badly on some processes, notably nonmimimum phase processes and those with complex instability.

13.6.2 Overcoming prediction mismatch by prediction stabilisation

The reader will not be surprised to hear that the major cause of poor performance is prediction mismatch and hence the removal of prediction mismatch is the most logical means of improving the performance of PFC. This was discussed at length in Chapters 6 and 7 and hence is not repeated here. Instead a summary is given.

. The main cause of prediction mismatch is that the open-loop predictions are divergent. Hence prediction mismatch can only be reduced significantly by working with stable predictions. This means that the d.o.f. commonly adopted in PFC (Section 13.2.4) must be modified. Possible means of doing this were given in Chapter 7. For instance, the predictions of (3.21) or (3.74) can be rewritten in transfer function form as

$$\underset{\rightarrow}{y}(z) = \frac{b(z)\Delta\underset{\rightarrow}{u}(z) + p(z)}{A(z)} \tag{13.11}$$

Hence one can parameterise (see Section 3.7.2) the future inputs that stabilise $\underset{\rightarrow}{y}(z)$. One can prestabilise in many different ways such as those illustrated in Section 7.3: (i) dead beat conditions; (ii) EUM conditions and (iii) optimal terminal conditions. In addition, one other option not discussed in Chapter 7 is the use of necessary and sufficient conditions [107, 133]; these are similar to EUM but do not restrict the future input trajectory to be FIR.

Remark 13.3 *Extensions of the prestabilisation approach to augment the constraint handling facility in PFC do exist [132] but are a little specialised for this book and hence omitted.*

Summary: The most obvious cause of poor performance in PFC is prediction mismatch. In the case of unstable processes, this can be alleviated a little by the use of prestabilised predictions.

13.6.3 Numerical examples

This section will demonstrate how the PFC algorithm, despite the use of only one coincidence point can still stabilise unstable processes, some with quite complex dynamics, and still give good performance. Moreover it will show the extent to which a global search on the coincidence horizon can be used for tuning purposes.

13.6.3.1 Link between tuning and the closed-loop poles

Plots are given in Figures 13.4-13.7 of the maximum modulus closed-loop pole against coincidence horizon for a number of different choices of Ψ and a number of examples given in Table 13.1. The reader should note that the closed-loop poles vary significantly with coincidence horizon. From the plots the reader can choose the coincidence horizon giving a closed-loop dynamic closest to Ψ. The notation in the plots is:

- PFC based on open-loop predictions (3.75) – dashed lines.

TABLE 13.1

Models and simulation parameters

Pole details	$G(z)$	Tuning parameters
Example 1 (pole at 1.5)	$\frac{z^{-1}-0.3z^{-2}}{1-1.9+z^{-1}0.48z^{-2}+0.18z^{-3}}$	$(n_y = 3, \Psi = 0.6)$ $(n_y = 10, \Psi = 0.6)$
Example 2 (pole at 1.49, zero at 1.22)	$\frac{0.2126z^{-1}-0.2594z^{-2}}{1-2.3967z^{-1}+1.3499z^{-2}}$	$(n_y = 7, \Psi = 0.5)$ $(n_y = 15, \Psi = 0.8)$
Example 3 (pole at 1.22, zero at 1.35)	$\frac{0.18z^{-1}-0.2432z^{-2}}{1-2.1262z^{-1}+1.1052z^{-2}}$	$(n_y = 6, \Psi = 0.5)$ $(n_y = 15, \Psi = 0.8)$
Example 4 (poles $1.2068 \pm 0.1885i$)	$\frac{0.2661z^{-1}-0.2172z^{-2}}{1-2.4136z^{-1}+1.4918z^{-2}}$	$(n_y = 2, \Psi = 0.6)$ $(n_y = 8, \Psi = 0.2)$

- PFC based on stabilised EUM predictions – dotted lines.

- PFC based on IHPC [107] predictions – solid lines

Open-loop predictions: The slowest pole always tends to one for large horizons. The link between the achievable closed-loop pole and Ψ is weak for many examples and good performance, if possible or stable at all, can usually only be obtained for small horizons.

EUM predictions: PFC was able to stabilise all the examples, but sometimes only for large horizons.

IHPC predictions: Pole behaviour was far more consistent for all coincidence horizons.

13.6.3.2   Closed-loop step responses

The closed-loop step responses analogous to Figures 13.4–13.7 are given in Figures 13.8–13.11 for sensibly chosen Ψ and coincidence horizon.

Open-loop predictions: For some examples good performance is possible, but only with small horizons. Other examples cannot be stabilised at all.

EUM predictions: These were able to stabilise all the examples, but sometimes only for large horizons.

IHPC predictions: Behaviour was far more reliable for all coincidence horizons.

Summary: Prestabilisation allowed a guarantee of the stability of PFC for a large enough horizon. Moreover the use of IHPC prestabilisation gave better reliability. However, for some models open-loop predictions worked well with small coincidence horizons.

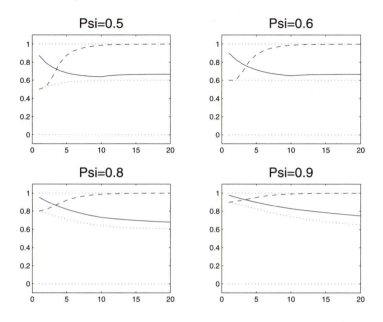

FIGURE 13.4

Closed-loop poles vs coincidence horizon for example 1.

13.7 Conclusions

PFC can be a very effective strategy despite the apparent simplicity of the underlying concepts. Where the problem concerned has fast sample rates it is an obvious option to consider first; that is, why consider a full scale, expensive and complicated MPC algorithm when a simple one is good enough.

Sometimes it is better to use a fast sampling rate (fast update of the receding horizon) with some prediction mismatch than to use slower sampling rate and less prediction mismatch. PFC allows the former, as it allows fast sampling rates. Moreover, due to the algorithm simplicity it is more straightforward to adapt for nonlinear models.

Summary: Despite its apparent simplicity, PFC often gives very good performance, with constraint handling, quite similar to that achievable with a far more complex MPC algorithm.

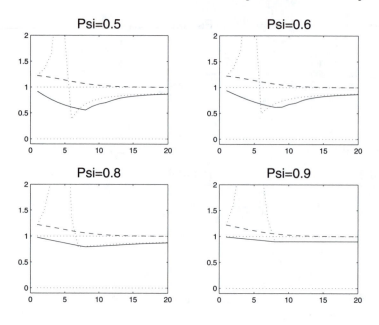

FIGURE 13.5

Closed-loop poles vs coincidence horizon for example 2.

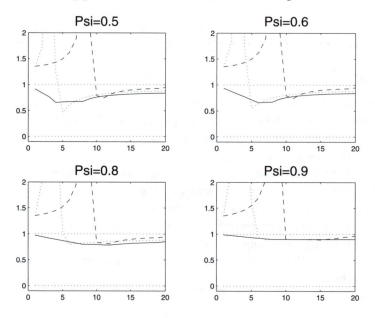

FIGURE 13.6

Closed-loop poles vs coincidence horizon for example 3.

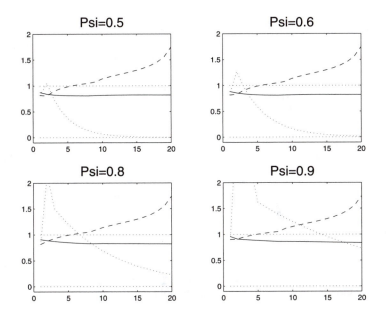

FIGURE 13.7

Closed-loop poles vs coincidence horizon for example 4.

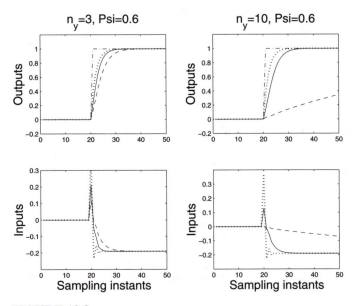

FIGURE 13.8

Unconstrained simulations for example 1.

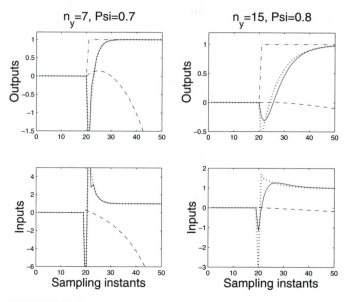

FIGURE 13.9

Unconstrained simulations for example 2.

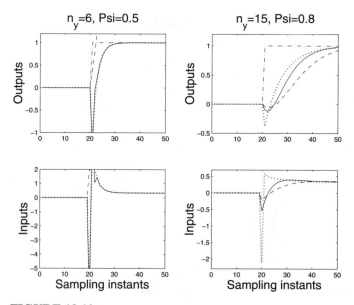

FIGURE 13.10

Unconstrained simulations for example 3.

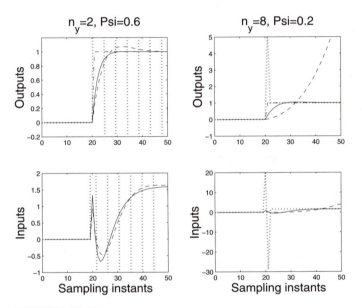

FIGURE 13.11

Unconstrained simulations for example 4.

14

Multirate systems

A book on predictive control would seem somewhat incomplete without some discussion of multirate (MR) systems simply because MPC is natural framework for handling models that may not fit into the usual linear setup. The reader will recall that the main building blocks of MPC are a prediction model, a performance index and some degrees of freedom. If one can formulate these, then the design of an MPC law can be seen as straightforward for many cases. As will be seen, although a MR system does not easily lend itself to conventional linear models, nevertheless one can supply a suitable prediction model and hence MPC design for this case is straightforward.

Summary: MPC forms a natural framework within which to handle multirate systems.

14.1 An introduction to multirate systems

A system is considered MR [28] when the inputs and outputs of a system are sampled at different rates. Typically this would be necessary if there were no restrictions on the speed at which the input is updated (denote this the fast rate (FR)), but output measurements are available only at a relatively slow rate (SR), for instance, where a laboratory test is needed. MR systems take many forms depending on the system dimensions and the sampling rates used; and although quite common in industry, such systems have recieved relatively little study from process control [74, 139] academics. The purpose of this chapter is to give a brief overview of some of the main concepts and to show how an MPC law might be developed.

In this book I have tried to avoid giving too many fine details where this does not add to understanding and as such this chapter will deal solely with SISO dual-rate systems; the algebra for a MIMO MR systems can be quite cumbersome [139] without adding much to the basic concepts. Hence, hereafter it is assumed that the output sample period is a simple multiple of the input sample period.

Summary: This chapter only gives an introduction to key concepts in MPC for MR systems.

14.2   Background on model and controller structure

MR systems cannot be modelled by conventional one step ahead difference models without substantial modification. This section gives a quick introduction to simple models that can be used for dual rate (DR) systems. These models are needed for analysis and design.

14.2.1   Modelling of MR systems

In single rate control, one assumes that an output measurement is available every sampling instant and hence one can use z-transforms and conventional linear theory to analyse the behaviour of the nominal loop. However, in a DR system this assumption breaks down; that is, it is not obvious how to use z-transforms and hence it is not immediately obvious how to analyse the model.

For instance, if the output is available only every m samples of the fast sample rate, a typical model might be:

$$y_{k+m|k} + a_1 y_k + a_2 y_{k-m} + \cdots + a_n y_{k-(n-1)m} = b_1 u_k + b_2 u_{k-1} + \cdots + b_n u_{k-n+1} \quad (14.1)$$

The model is based on the output every mth sample (called the SR) and the input every sample (the FR). Any analysis needs to take proper account of the presence of both these two distinct sampling rates. For convenience this chapter assumes that the slower sample period is a integer multiple of the faster.

14.2.2   Control trajectory update

The second dynamic component in the loop is the controller. If the output is updated only every mth sample, then the control law can only be updated every mth sample; and hence the control law is implicity operating at the slow sample rate (even with inferential control) – it does not have a linear fast rate equivalent. In fact it is well known [139] that the control law is equivalent to a periodically *time varying* FR control law. So symbolically there are m different implied laws:

$$u_{k+i|k} = f_i(\underleftarrow{u}, \underleftarrow{y}, \underrightarrow{r}), \quad i = 0, 1, \ldots, m-1 \quad (14.2)$$

where $\underleftarrow{u}$ has past inputs every sampling point whereas $\underleftarrow{y}$ comprises past outputs only at the slow sampling points. Although the control law is time varying, it varies periodically over each m samples and therefore can be represented by a linear equivalent at the slow sample rate.

Summary:

- Linear analysis cannot be applied at the fast sample rate because the control law is time varying.

- At the slow sample rate, linear analysis does apply.

14.2.3 Control law structure in MR systems

An important question to be answered next is: *how is the time varying controller (14.2) to be determined?* To be more precise, how is one to update the control action in between the output measurements, that is, $u_{k+i|k}$, $i = 1, ..., m-1$? At these points there is no new output information on which to base an updated optimisation.

There are two popular solutions to handling the MR nature of the process which then determine the controller structure:

1. Inferential control (e.g. [72])

2. Lifting [68]

These will be introduced next.

14.2.3.1 Inferential control (IC)

IC makes use of an internal process model (see Figure 14.1) which operates at the FR. This model is used to supply output estimates at the fast sample rate. The controller is based on the model output, not the process output and hence can operate at the fast sample rate.

This model can be viewed as analogous to a state estimator which supplies state values to be used in lieu of the actual (and unknown) state. Obviously some correction is needed and this is provided using a DMC mechanism (Section 3.5), that is, a simple disturbance model for the offset between the model and process outputs. This offset estimate can be updated only every mth sample.

However, this approach needs more study as there are several obvious weaknesses:

1. The state estimator/internal model receives actual output updates very slowly and this could have repercussions on accuracy.

2. The approach relies on knowledge of a fast SR model which would have to be identified from MR data; recent work [74] has shown that this is possible in some cases but a clear understanding of the robustness of these models constitutes work in progress.

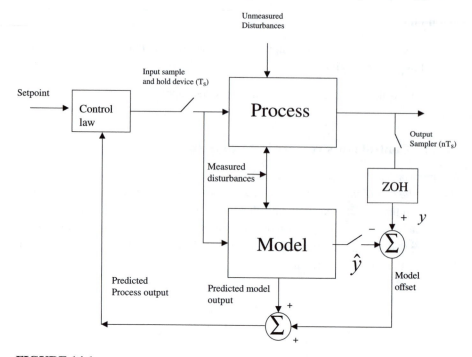

FIGURE 14.1
Internal model structure.

14.2.3.2 Lifting

A more popular alternative [68] has been to use lifting. In essence this transforms a MR SISO system to a single rate MIMO system or if the system were already MIMO it increases the dimension.

In simple terms the inter sample inputs u_{k+i}, $i = 0, ..., m - 1$ are treated as independent variables, so a SISO system at the FR is transformed to a $1 \times m$ MISO system at the SR. An illustration will make the procedure clearer.

Consider a FR state-space model of the form

$$x_{k+1} = Ax_k + Bu_k; \quad y_k = Cx_k \tag{14.3}$$

The DR equivalent* (SR model) to this system could be written down as

$$x_{k+n} = \Gamma x_k + \Theta U_k; \quad y_k = C x_k; \quad U_k = \begin{bmatrix} u_k \\ \vdots \\ u_{k+m-1} \end{bmatrix} \tag{14.4}$$

$$\Gamma = A^m, \quad \Theta = \begin{bmatrix} A^{m-1}B & \cdots & AB & B \end{bmatrix}$$

Model (14.4) will be denoted the lifted model, as the input has been lifted from u_k to U_k. Also the output/state is updated only every n samples of the FR. Effectively this gives a SR model with a lifted input. In many scenarios [74] one may be able to identify Γ, Θ (or equivalent model form) from input/output data fairly easily but not A, B.

As the lifted system is SR (the SR at which the output is updated) one can use linear design and analysis methods. However:

1. There is the price of working in a significantly increased dimension and hence the design itself may be far more complex.

2. There is the so called causality constraint [17, 140] whereby one needs to ensure that the structure of the controller does not make current controls dependent on future controls. This implies a structure constraint that the feedthrough term in the controller must be block lower triangular.

3. For both IC and lifting based schemes there is also the issue of intersample ripple [147]; to avoid this one must place additional constraints in the controller structure.

Summary:

- IC assumes knowledge of the FR model and hence forms a linear control law at the fast rate FR but has an estimator running at the SR.

- Lifted control forms a linear controller in an augmented space at the SR but is applicable when a FR model is unknown.

14.2.4 Overview of chapter

This chapter will illustrate and contrast the use of IC and lifted control methodologies within MPC. This will be done by showing how an MPC control law can be formulated, illustrating the resulting closed-loop responses and giving some insight into the respective advantages and disadvantages.

---

*Lifting can be applied to systems where many different sample rates apply and hence is a powerful tool but the notation is cumbersome so is omitted here.

14.3   GPC (finite output horizon) controllers for MR systems

This section will illustrate how a MR control law can be developed in conjunction both with an IM and with lifting. It is assumed that the reader is by now fluent in constructing prediction models and hence the focus is on the key steps rather than on the mathematical details. This section will overview the components required which as usual are: (i) a performance index; (ii) a prediction model and (iii) a parameterisation for the d.o.f. For simplicity of notation only the SISO case is given, the extension to the MIMO case being automatic along the lines of that given in Chapter 4.

14.3.1   The internal model/inferential control

The performance index to be used here is the usual one, for instance, that of eqn.(4.4) or (7.18) as appropriate. Given the use of an IM, the predictions are given from equations such as (3.63) or (3.64). The only major difference is that the offset correction term in the predictions is the most recent available (see Figure 14.1), so it is only updated every m-samples.

Given the above components, the predictive control law takes the appropriate form of (4.70 or 14.5) but where the terms entering through M_k are updated at the slow sample rate only. The derivation of this is straightforward and left to the reader.

$$\Delta \mathbf{u}_k = \underbrace{P_r \underset{\rightarrow}{\mathbf{r}} - \check{D}_k \underset{\leftarrow}{\Delta \mathbf{u}} - \check{N}_k \underset{\leftarrow}{\hat{\mathbf{y}}}}_{\text{Fast update}} - \underbrace{M_k \mathbf{y} + M_k \hat{\mathbf{y}}}_{\text{Slow update}} \qquad (14.5)$$

It is noted that in this case one can use infinite horizons, the closed-loop paradigm and other variations on the nominal control law without any particular increase in complexity.

> **Summary:** The IC law comprises a linear part dependent on the model states that operates at the FR and a separate component dependent on the offset between the process measurement and model output which is updated at the SR.

14.3.2   MPC in the lifted environment

This case is slightly different, as a FR model is assumed unknown; hence model (14.4) is used for prediction. The implications[†] are that the performance index is

---

[†]This performance index is consistent with the assumptions made about state-space implementations in Chapter 4 , e.g. eqn.(4.61).

reformulated to only cost every mth output, as the intersample outputs cannot be estimated. Hence:

$$\min_{u_i,\ i=0,\dots,n_c-1} J = \sum_{i=1}^{n_y} (r-y_{k+mi})^2 + \sum_{i=0}^{n_c-1} \lambda \left(u_{k+i}-u_{ss}\right)^2$$

$$s.t. \begin{cases} u_{k+i}=u_{ss},\ i \ge n_c \\ \text{constraints} \end{cases} \tag{14.6}$$

The corresponding prediction vectors are:

$$\underset{\rightarrow}{y} = \begin{bmatrix} y_{k+m} \\ y_{k+2m} \\ \vdots \\ y_{k+n_ym} \end{bmatrix};\ \underset{\rightarrow}{u} = \begin{bmatrix} \underset{\rightarrow}{u}_1 \\ Z \end{bmatrix};\ \underset{\rightarrow}{u}_1 = \begin{bmatrix} u_k \\ u_{k+1} \\ \vdots \\ u_{k+n_c-1} \end{bmatrix} \tag{14.7}$$

where Z is a vector of zeros. It is assumed that the input can be updated every sample. Although the model is in terms of lifted variable U_k, the predictions have been stated in terms of u_k. This is because the number of d.o.f. in the optimisation relates to the dimension of $\underset{\rightarrow}{u}$ and need not be a linear multiple of m, which could be implied by a naive use of $\underset{\rightarrow}{U}$. The relationships are obvious and left for the reader.

Following the lines of Section 3.2, but using (14.4, 14.7), the prediction model takes the form

$$\underset{\rightarrow}{y} = \underbrace{[H_1|H_2]}_{H} \begin{bmatrix} \underset{\rightarrow}{u}_1 \\ Z \end{bmatrix} + Px_k$$

$$H = \begin{bmatrix} \Theta\ 0\ 0\dots \\ \Gamma\Theta\ \Theta\ 0\dots \\ \vdots\ \vdots\ \vdots\ \vdots \end{bmatrix};\ P = \begin{bmatrix} \Gamma \\ \Gamma^2 \\ \vdots \end{bmatrix} \tag{14.8}$$

where the partition of H is conformal with that of $\underset{\rightarrow}{u}$ and hence does not need to be conformal with a neat partition of $\underset{\rightarrow}{U}$. One can now substitute this prediction into (14.6) to derive the first n_c steps of the optimal control trajectory as:

$$\underset{\rightarrow}{u}_1 - Lu_{ss} = \underbrace{[H_1^T H_1 + \lambda I]H_1^T P(x-x_{ss})}_{K_{mr}} \tag{14.9}$$

where L is an n_c vector of ones and u_{ss}, x_{ss} depend upon r and the offset term. The corresponding $\underset{\rightarrow}{U}$ can be constructed from this if required.

Ignoring u_{ss}, x_{ss}, the control law would be given as the first m terms of

$$\underset{\rightarrow}{u}_1 = -K_{mr}x \quad \Rightarrow \quad \begin{bmatrix} u_{k|k} \\ u_{k+1|k} \\ \vdots \end{bmatrix} = \underbrace{\begin{bmatrix} K_1 \\ K_2 \\ \vdots \end{bmatrix}}_{K_{mr}} x \tag{14.10}$$

where in general $K_1 \neq K_2 \neq K_2, \ldots$. Hence, the control law is periodically time varying.

Summary: Lifting gives rise to a periodically time varying control law. The control trajectories are updated only every m samples and hence one has to live with any prediction mismatch in the optimal predictions for m samples before an update can take place.

14.3.3 Comparison of IC and lifting based control schemes

One of the main factors in the success of MPC is the receding horizon. That is, even if the class of predictions is simplistic (such as GPC with $n_u = 1$) and carries significant prediction mismatch, neverthless reasonable performance can still follow. This is because the decisions are continually updated and at each sampling instant a new d.o.f. is included to allow further improvement of the predictions. These benefits still apply to the IC case.

However, lifting based control only updates decisions every m samples, at which point n_c new d.o.f. are introduced. As a consequence one has to live with any prediction mismatch for m samples before improvements can be made and hence intersample (between slow samples) behaviour can be poor unless a large number of d.o.f. are used[‡]. To put some focus on this [136], the assumption that in the predictions $u_{k+i} = u_{ss}$ can be catastrophic if $n_c < m$, as this assumption will actually be implemented.

A second major weakness of lifting based methods is that intersample behaviour is unobserved and hence uncontrolled [147]. This may be a structural constraint if the ouputs are not observable. However, where a FR model can be reliably identified and hence IC can be utilised, then inter sample outputs can be inferred and hence should be controlled.

Summary: Lifting based control suffers from a slow receding horizon update and hence, with small n_c as typical in real applications, performance could be poor. Moreover there is no control of intersample behaviour. IC avoids this problem and hence should give better performance in general. However, it requires the assumption that a FR model can be identified.

[‡]The reader will recall that high n_c reduces prediction mismatch.

TABLE 14.1

Closed-loop run time costs

J

Lifted algorithm	3.18
Inferential control	2.21

14.4 Simulation example contrasting lifting and IC

For simplicity this section compares lifting and IC algorithms with the cost of (14.6), that is with a finite output horizon and costing only every mth output. Similar conclusions would apply for the infinite horizon (dual mode) case.

14.4.1 Simulation details

Consider an example with a FR state-space model

$$x_{k+1} = \begin{bmatrix} 0.3 & 0.5 \\ 0.1 & 0.9 \end{bmatrix} x_k + \begin{bmatrix} 0.1 \\ 0.2 \end{bmatrix} u_k; \quad y_k = [1 \ 0]x_k \qquad (14.11)$$

For the lifted algorithm one would assume that only the equivalent model of form (14.4) is known. Assume that the output is sampled 4 times slower than the input, i.e. $m = 4$. The control laws of (14.9) and (14.5) are implemented for $n_c = 1$ with $n_y = 8$, $\lambda = 1$. The simulations are displayed in Figure 14.2 for the outputs and inputs; circles and dotted lines are used for the IC algorithm and crosses and solid lines are used for the lifted algorithm. The x-axis has units of the fast sample rate so new output measurements are given only every fourth sample. The corresponding closed-loop run time costs are given in Table 14.1 for $n_c = 1$.

14.4.2 Discussion of simulations

It is clear both from the table and the figure that the use of a fast receding horizon update allowed in IC has given a dramatic improvement in performance, even though there has been no new output measurements. The limitation of the prediction assumption in the lifted algorithm is very clear in Figure 14.2 where it can be seen that the input moves to a *poor* default value, that is u_{ss}, during the later intersample periods. If one uses a fast receding horizon [135], the negative effects of this poor assumption can be alleviated, as the only predicted value actually implemented is the current and the far future is continually updated. In the lifted framework, the first m moves are used and hence one is forced to use a poorly defined input trajectory; that is, to implement a poorly defined prediction.

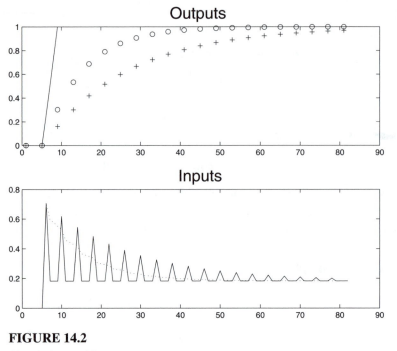

FIGURE 14.2

Simulations with $n_c = 1$.

14.4.3   Summary of comparison

Finite horizon algorithms typically use input predictions which do not match the
expected or desired closed-loop behaviour. This limitation is partially overcome by
the use of the receding horizon concept whereby one updates the predictions at every
sample instant so that there is a continual improvement on the initial assumption.
Unfortunately in a lifted framework, the receding horizon update only takes place at
a SR (every m samples) and as a consequence a naive use of finite horizon MPC will
cause the control law to inherit a poor input prediction. One obvious solution to this
is to use IC, which was popular in some early papers on MR systems [72]. IC allows
the use of a fast receding horizon update to improve performance. However it should
be emphasised that IC assumes the knowledge of a FR model which is not always
a realistic assumption. Alternative ways around this are a topic of current research
[136].

Summary:

1. IC will usually outperform lifting based algorithms when the same number of d.o.f. are deployed.

2. IC requires knowledge of a FR model which lifting based approaches do not.

3. Lifting algorithms can fail catastrophically due to prediction mismatch if not set up wisely [136].

14.5 Infinite horizons in the MR environment

This book demonstrated in Chapter 6 that in general one could reduce prediction mismatch and hence improve performance, for the same number of d.o.f., by using dual mode algorithms and infinite horizons. Clearly the same statement applies to MR systems.

In an ideal situation one wants the optimised open-loop predictions to match the actual closed-loop behaviour, such as in LQMPC. Hence the optimisation is well posed. The consequence of the change to LQMPC is that the input discontinuities apparent in Figure 14.2 should not occur, even in the lifted environment. To rephrase this, in the nominal case, the optimal input trajectory at time k will match exactly the optimum computed at the previous sample (in the absence of constraints). Hence whether one updates the control law at the FR or the SR, the control inputs will be the same (for the same performance index).

This is easy to demonstrate and hence is omitted here. But it raises the confusing question of should IC still be prefered to lifting based control?

The equivalence between lifting and IC made two large assumptions:

1. The performance index J was the same for both. That is, they costed the outputs only at the slow sample times.

2. In the nominal case, the optimal trajectories did not change from one sample to another.

Of course both of these assumptions do not hold in general.

1. If a FR model were available, then one can infer and therefore should cost intermediate outputs. Therefore to use the same performance index for both algorithms is not sensible. Logically the lifted approach cannot not give as tight control over the unmeasured and hence uncontrolled intersample outputs [147].

2. The optimal predicted trajectories are rarely globally optimal due to the restriction to n_c d.o.f. and constraint handling. Hence as one adds further d.o.f. at each sample, so a change in the proposed input trajectory is likely to occur.

Summary: Even with infinite horizon algorithms, IC should outperform lifting based control, as it both: (i) observes (controls) intermediate outputs and (ii) makes used of a faster receding horizon (but see [136]).

14.6   Conclusions

It was shown that the assumption, usual with finite horizon algorithms, that the predicted input move to a fixed value after n_c steps does not mesh well with MR control design, especially where n_c is small as typical in practice. This is because the assumption is made to reduce computation and not to improve control and does not match expected closed-loop behaviour well enough (prediction mismatch). Good control is recovered only by applying the receding horizon concept at a fast enough update rate and this is not possible within conventional lifting based methods.

Infinite horizon algorithms overcome this limitation as they are set up to ensure a good match between predictions and expected closed-loop behaviour. Hence in this case the lifting based algorithm gives only a small deterioration in performance compared to a FR equivalent.

The faster the rate of receding horizon update the more quickly extra d.o.f. can be introduced to improve performance. Hence IC will always outperform lifting during constraint handling, even for the infinite horizon case. However, FR models are not always available. Moreover work in progress [136] is looking at means of obtaining a fast receding horizon update with a slow (lifted) model.

Summary: Predictive control can be applied easily to the MR case. However the best means of doing this depends upon what models can be inferred. This topic is still in relative infancy and deserves far more attention from the academic community.

15

Modelling for predictive control

This chapter is deliberately brief, as it has the purpose of introducing the concept only. The topic, that is, how does one set up an identification to support an MPC design, is currently an active area of research. The basic premise taken is that the modelling stage should have a synergy with the use of the model in the control design; they are not distinct tasks. Therefore this chapter considers how one's approach to modelling changes when one has a particular control law in mind. Here the identification is being used alongside a predictive controller.

Summary: The identification method adopted should support the intended control design.

15.1   Introduction

Identification is the process whereby one discerns a system model from available information [76]. Model identification could be considered as an end in itself; however, often the model is wanted as a base for control law design and hence the potential utility of the model for this purpose must be considered during the identification. The basic premise [126] taken in this chapter is that there should be a synergy between the model identification and the role of the model in control law design. Such a view point differs from typical practise within MPC whereby one identifies a model via some algorithm and then uses it without considering possible interactions in these two steps. In the case of predictive control, the potential for synergy is large and should be exploited.

15.1.1   Multi-models in MPC

MPC is based on system predictions and in particular how these predictions depend upon the current process state and future control moves. As discussed in Chapter 6 there is a strong relationship between prediction accuracy and performance: there is a need for the predictions used in the optimisation to be a good match to the actual behaviour, as a significant mismatch between predicted and actual behaviour could make the optimisation meaningless.

In [141, 142] it was noticed that conventional one-step ahead prediction models arising from least squares identification (LSI) routines often gave quite poor predictors. The solution proposed was to filter the data before doing the identification so to emphasise the slower frequencies which are more important in long range prediction. However, any filtering trades off one frequency range for another, which effectively trades off accuracy in the near horizon against accuracy in the far horizon. Moreover, extensions to the MIMO case, although possible in principle, were not straightforward and it was stated that in practice a sensible low-pass filter often was equally effective; this is still the most common procedure in industry.

An alternative method to the use of a single model is to identify a separate model for each prediction horizon and hence not to rely on a single model capturing both the fast and slow dynamics. Here this is termed a *multi-model*. This underlying concept was considered as early as the 1980's within MUSMAR algorithm [39] and also in [27, 75, 120]. More recently [126] gave some evidence of the efficacy of the so called multi-model approach within MPC and hence this chapter will give a brief overview of the approach.

Remark 15.1 *Some authors have looked at how to use optimised filters along with recursive relationships to miniminse sensitivity of predictions to uncertainty (e.g. [43, 44, 148]). However, that work presupposes knowledge of the plant and the uncertainty and hence is more relevant to the discussions in Chapter 9.*

Summary: This chapter will give an overview of the multi-model approach to modelling, as this gives some synergy with MPC.

15.1.2   Feasibility issues

Although not illustrated in this chapter, one of the main reasons that predictions need to be accurate is for constraint handling, in particular for state constraints and terminal constraints. One assumes that one can test for constraint satisfaction of the system evolution by ensuring the predictions satisfy constraints. However, if there are significant errors in the predictions, then this assumption will break down. As mentioned in Chapter 8, the use of infeasible trajectories (due to a mismatch between the predictions and the reality) can lead to disaster in the worst case and poor performance in many cases.

Methods do exist for ensuring constraint satisfaction given some uncertainty in the predictions, but these methods tend to be conservative in the assumptions they take and would probably not be favoured in practice.

Summary: For reliable and robust constraint handling, the predictions must be as accurate as possible.

15.1.3 Closed-loop identification and iterative feedback tuning

Recent work has illustrated how closed-loop identification (e.g. [46, 47, 70]) can support the control design more closely. This work is both mathematically complex and can have slow convergence rates and hence as it sits more in the topic of identification, only a brief summary will be given here.

The basic argument is that if one identifies in the closed-loop, then the implied frequency content of the data puts greatest emphasis on critical frequencies, that is, where $1 + GK$ is smallest. Or in other words frequencies near the gain/phase cross over frequencies. Hence one expects better modelling accuracy in frequency ranges where this is critical for performance. This in turn will result in a model based design which is less sensitive to modelling errors, as the modelling errors have been minimised where they have the greatest effect on sensitivity.

Another developing area [48, 49] is that of iterative feedback tuning (IFT). In this, one does not form a model at all but instead goes directly from measured process data to the control design. By missing out the intermediate step, less bias (errors) should be built in and hence a better control design can result. However, these methods are very demanding on quantities of measured data* and would usually be implemented on-line in an adaptive sense. This means there are two *major* weaknesses when used in conjunction with MPC:

1. The data requirements for the MIMO case can be prohibitive.

2. The methods are only applicable in the constraint free case.

> **Summary:** This book does not look at closed-loop identification or IFT. However, as they are becoming increasingly accepted in the literature, the reader may want to investigate them.

15.2 Predictions models

This chapter uses matrix fraction description (MFD) models but in fact there is a significant overlap with subspace modelling techniques. The notation to be adopted matches that given in Section 3.4. Hence the notation for the mth order one-step ahead MFD model used in this chapter is as follows:

$$\mathbf{y}_{k+1|k} = N_1^{[1]}\Delta\mathbf{u}_k + N_2^{[1]}\Delta\mathbf{u}_{k-1} + \cdots + N_l^{[1]}\Delta\mathbf{u}_{k-l+1} \\ -D_1^{[1]}\mathbf{y}_k - D_2^{[1]}\mathbf{y}_{k-1} - \cdots - D_m^{[1]}\mathbf{y}_{k-m+1} \tag{15.1}$$

---

*A similar method [124], that is, modelling the performance index directly from data, also illustrated this problem.

Matrices $D_i^{[1]}$, $N_i^{[1]}$ are the model parameters and the superscript $(.)^{[j]}$ is used to denote model parameters for j-step ahead prediction. Hence the n-step ahead predictions[†] are given from

$$\mathbf{y}_{k+n|k} = M^{[n]}\Delta\mathbf{u}_k + M^{[n-1]}\Delta\mathbf{u}_{k+1|k} + \ldots + M^{[1]}\Delta\mathbf{u}_{k+n-1|k}$$
$$+ N_2^{[n]}\Delta\mathbf{u}_{k+1} + \ldots + N_l^{[n]}\Delta\mathbf{u}_{k-l+1} \quad (15.2)$$
$$- D_1^{[n]}\mathbf{y}_k - D_2^{[n]}\mathbf{y}_{k-1} - \ldots - D_m^{[n]}\mathbf{y}_{k-m+1}$$

A general prediction model is

$$\underrightarrow{\mathbf{y}} = H\underrightarrow{\Delta\mathbf{u}} + P\underleftarrow{\Delta\mathbf{u}} + Q\underleftarrow{\mathbf{y}}$$

$$H = \begin{bmatrix} M^{[1]} & 0 & 0 & \ldots \\ M^{[2]} & M^{[1]} & 0 & \ldots \\ M^{[3]} & M^{[2]} & M^{[1]} & \ldots \\ \vdots & \vdots & \vdots & \vdots \\ M^{[n]} & M^{[n-1]} & M^{[n-2]} & \ldots \end{bmatrix} ; \quad P = \begin{bmatrix} N_2^{[1]} & \ldots & N_l^{[1]} \\ N_2^{[2]} & \ldots & N_l^{[2]} \\ N_2^{[3]} & \ldots & N_l^{[3]} \\ \vdots & \vdots & \vdots \\ N_1^{[n]} & \ldots & N_l^{[n]} \end{bmatrix} ; \quad Q = \begin{bmatrix} D_1^{[1]} & \ldots & D_m^{[1]} \\ D_1^{[2]} & \ldots & D_m^{[2]} \\ D_1^{[3]} & \ldots & D_m^{[3]} \\ \vdots & \vdots & \vdots \\ D_1^{[n]} & \ldots & D_m^{[n]} \end{bmatrix}$$

$$(15.3)$$

The question that arises then is, can we get a prediction model (15.3) which gives more accurate prediction by trying to identify the prediction matrices H, P, Q directly rather than via the recursive relationships of (3.51) on model (15.1)? The answer is yes and this is described next.

Summary: The MPC optimisation depends upon the prediction equation (15.3) and not the underlying MFD model (15.1). For an accurate control law, the determination of matrices H, P, Q is far more important than an accurate knowledge of (15.1).

15.3   Identifying a multi-model

Predictive control requires knowledge of the parameters of eqn.(15.2), $n = 1, 2, \ldots, n_y$ as these comprise the components of matrices H, P, Q. It is straightforward to use *independent* least squares algorithms to compute the matrices $N_i^{[n]}, D_i^{[n]}, M_i^{[n]}$ for all i, n; that is, to find the best predictor first for one-step ahead prediction, secondly for two-step ahead prediction, etc. As the predictors are identified independently, it is not necessary that recursive relationships (3.51) hold. In fact, forcing relationships (3.51) to hold may make the predictions less accurate when a model is formed from real data.

[†]These have been called ARMarkov models [54].

15.3.1 Algorithm details

In order to keep the dimension of the optimisations to a minimum, one should compute only one row of the matrices H, P, Q at a time. For example, dropping the superscipts to improve clarity, rewrite the generic n-step ahead prediction equation of (15.2) as

$$
\begin{aligned}
y_{k+n|k} = {} & M_n \Delta u_k + M_{n-1} \Delta u_{k+1|k} + \cdots + M_1 \Delta u_{k+n-1|k} \\
& + N_2 \Delta u_{k+1} + \cdots + N_l \Delta u_{k-l} \\
& - D_1 y_k - D_2 y_{k-1} - \cdots - D_m y_{k-m}
\end{aligned} \tag{15.4}
$$

Next, introduce the notation $(.)^{(i)}$ to denote the operation of taking only the ith row, e.g.

$$
y_{k+n|k} = \begin{bmatrix} y_{k+n|k}^{(1)} \\ y_{k+n|k}^{(2)} \\ \vdots \\ y_{k+n|k}^{(s_y)} \end{bmatrix} ; \quad
M_j = \begin{bmatrix} \mathbf{m}_j^{(1)} \\ \mathbf{m}_j^{(2)} \\ \vdots \\ \mathbf{m}_j^{(s_y)} \end{bmatrix} ; \quad
N_j = \begin{bmatrix} \mathbf{n}_j^{(1)} \\ \mathbf{n}_j^{(2)} \\ \vdots \\ \mathbf{n}_j^{(s_y)} \end{bmatrix} ; \quad
D_j = \begin{bmatrix} \mathbf{d}_j^{(1)} \\ \mathbf{d}_j^{(2)} \\ \vdots \\ \mathbf{d}_j^{(s_y)} \end{bmatrix} \tag{15.5}
$$

Now the ith row of (15.4) can be written as

$$
\begin{aligned}
y_{k+n|k}^{(i)} = {} & \mathbf{m}_n^{(i)} \Delta u_k + \mathbf{m}_{n-1}^{(i)} \Delta u_{k+1|k} + \cdots + \mathbf{m}_n^{(i)} \Delta u_{k+n-1|k} \\
& + \mathbf{n}_2^{(i)} \Delta u_{k+1} + \cdots + \mathbf{n}_l^{(i)} \Delta u_{k-l+1} \\
& - \mathbf{d}_1^{(i)} y_k - \mathbf{d}_2^{(i)} y_{k-1} - \cdots - \mathbf{d}_m^{(i)} y_{k-m+1}
\end{aligned} \tag{15.6}
$$

Now putting (15.6) in the format of a vector of model parameters multiplying a vector of model data, the n-step ahead predictor at sampling instant k can be represented as

$$
y_{k+n|k}^{(i)} = \mathbf{g}_{k,n}^T \theta_n^{(i)}; \quad
\mathbf{g}_{k,n} = \begin{bmatrix} \Delta u_k \\ \Delta u_{k+1|k} \\ \vdots \\ \Delta u_{k+n-1|k} \\ \Delta u_{k+1} \\ \vdots \\ \Delta u_{k-l+1} \\ y_k \\ y_{k-1} \\ \vdots \\ y_{k-m+1} \end{bmatrix} ; \quad
\theta_n^{(i)} = \begin{bmatrix} \mathbf{m}_n^{(i)T} \\ \vdots \\ \mathbf{m}_1^{(i)T} \\ \mathbf{n}_2^{(i)T} \\ \vdots \\ \mathbf{n}_l^{(i)T} \\ -\mathbf{d}_1^{(i)T} \\ \vdots \\ -\mathbf{d}_m^{(i)T} \end{bmatrix} \tag{15.7}
$$

One can then stack eqn.(15.7) over several sampling instants in the usual fashion, e.g.

$$\mathbf{v}_n^{(i)} = G_n \theta_n^{(i)}; \quad \mathbf{v}_n^{(i)} = \begin{bmatrix} y_{k+n|k}^{(i)} \\ y_{k+1+n|k+1}^{(i)} \\ \vdots \\ y_{k+r+n|k+r}^{(i)} \end{bmatrix}; \quad G_n = \begin{bmatrix} \mathbf{g}_{k,n}^T \\ \mathbf{g}_{k+1,n}^T \\ \vdots \\ \mathbf{g}_{k+r,n}^T \end{bmatrix} \qquad (15.8)$$

A typical least squares solution for $\theta_n^{(i)}$ is then given as

$$\theta_n^{(i)} = (G_n^T G_n)^{-1} G_n^T \mathbf{v}_n^{(i)} \qquad (15.9)$$

It should be noted that for any given prediction horizon, the matrix G_n is the same for each loop, (that is index 'i' above) and therefore the computationally demanding part of (15.9) need only be done once for each horizon.

The major drawback of the approach above is that for large prediction horizons, far more plant data will be required as the dimension of $\theta_n^{(i)}$ grows with n. In fact it can be shown that this becomes equivalent to identifying the step response directly and hence is no more demanding than methods currently adopted in industry. A similar technique calls these models ARMmarkov models [54].

> **Summary:** The multi-model identification algorithm consists of independent least squares problems which compute each row of H, P, Q in turn.

15.3.2 Including a noise model in the multi-model

It is well known in the identification literature [76] that in order to get an accurate model, then one needs specific assumptions on the noise model. The use of an incorrect noise model will give a bias in the model parameters. Research is currently ongoing in Edmonton, Canada to translate these results to the multi-model case and hence to show how a noise model can be incorporated efficiently. For now the reader should be aware that significant improvements in accuracy may be possible in some cases. This chapter introduces the principle concept of utilising the multi-model but does not focus on more advanced details of the identification algorithm.

15.3.3 Over- and underparameterisation

Usually it is dangerous to overparameterise a model as one can produce near cancelling pole zero pairs. If these pairs are unstable, then the model will give erroneous predictions even for small prediction horizons. However, with the multi-model, one computes the parameters of the prediction models independently for each prediction horizon. Therefore, the location and presence of implicit poles and zeros are of no consequence and one can overparameterise the model with some impunity, although one may have to use singular value decompositions or other numerically reliable techniques for solving (15.9).

In a similar vein, one can also underparameterise more easily. There may be cases where for a high order process the transient dynamics can be captured with a low order model and the asymptotic behaviour can be captured by a separate low order model. In this case one may find that one can capture the whole dynamic with a low order multi-model, that is, one where matrices P and Q have less columns than implied by the underlying full order process model. This would have advantages where a low order controller was required, as the controller order is related directly to the number of columns in P, Q (see Section 4.3.2) and also where there are limitations on the quantity of data.

Summary: The multi-model is far less sensitive to parameterisation errors than prediction models based on one-step ahead models.

15.4   Examples

In this section the efficacy of the multi-model is illustrated by way of some examples.

15.4.1   No parameterisation errors

In the ideal case of linear plant of known order and a suitable noise model, one would expect the one-step ahead model to be a near exact match (with sufficient data) to the true plant and hence there would be little to gain with a multi-model. This observation is borne out on examples.

15.4.2   Examples with parameterisation errors

In practice systems are never purely linear and neither are their orders known precisely and this is where the multi-model can offer improvements in prediction accuracy. Here two examples [126] are illustrated:

1. SISO model with two zeros and three poles.

2. High order nonlinear MIMO model with two outputs.

The models are simulated open-loop for a pseudo random binary sequence (PRBS) input and with a slow PRBS disturbance signal plus noise on the output. The data is then used to identify a one step ahead model (15.1) and a multi-model. Then, using an independent set of open-loop data, the standard deviation of the associated prediction errors for each model, each horizon and each output are computed. The

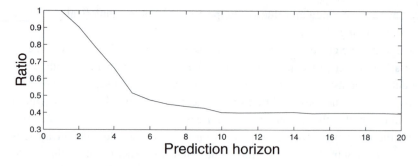

FIGURE 15.1

Ratio multi-model:one-step ahead model of the standard deviations of n-step ahead prediction errors.

results are displayed in Figures 15.1, 15.2 as the ratio of the errors (y-axis) for the multi-model and one-step ahead model versus prediction horizon (x-axis); any numbers less than one indicate that the multi-model has better modelling accuracy. The variables l, m are defined in eqn.(15.1). Figure 15.1 shows the results for model 1 using $l = 2$, $m = 2$, therefore underparameterising the number of poles. Figure 15.2 shows the results for model 2 with $l = 5$, $m = 5$. This is severely underparameterised. A log scale is required due to the total failure of the one-step ahead model for large horizons.

The observations are:

- The multi-model consistently gives significantly smaller standard deviations than one-step ahead models.

- For model 2, standard models cannot be used to form a useful prediction model [131]. The rapid deterioration of prediction errors with horizon is very fast (see Figure 15.2).

- It has also been observed that the inevitable degredation of prediction errors with horizon length can be slower with multi-models [126].

Summary: In the case of underparameterised models the multi-model can sometimes give far more accurate predictions.

15.5   Conclusions

This chapter has introduced the multi-model approach to modelling for predictive control and demonstrated its potential advantages over the more usual one-step ahead

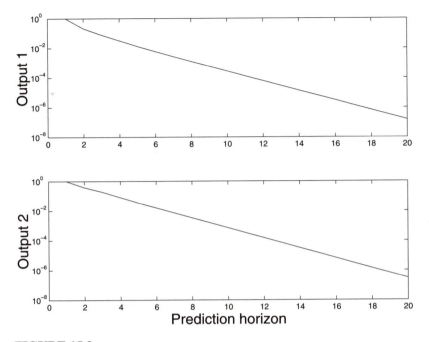

FIGURE 15.2

Ratio multi-model:one-step ahead model of the standard deviation of n-step ahead prediction errors.

prediction models.

It seems to be more robust to assumptions on model order. Often the model order is unknown and/or there is model mismatch due to nonlinearities. Therefore a technique which is robust to such errors is advantageous. The multi-model approach gives better accuracy when the order is underparameterised and when overparameterised, does not suffer from the near cancelling pole/zero effects that predictions based on one-step ahead models suffer from. One might also conjecture that the multi-model approach will lend itself much more readily to the modelling of multirate and stiff systems, in essence because one has much more freedom with its structure than a conventional one-step ahead prediction model. These issues are also discussed in [54].

Summary: Not much research has gone into exploring the full potential of multi-models and hence this is still a relatively open field. Clearly there are strong overlaps with subspace identification and potential links with IFT and closed-loop identification.

16

Conclusion

This book only scratches the surface of available knowledge and algorithms within linear MPC, let alone nonlinear MPC. The reader may be left wanting more illustrations, especially MIMO ones, and wondering why certain topics were omitted.

The simple answer is that I have tried to focus on understanding how the MPC algorithm works (or fails), giving the reader insight into the algorithms presented rather than copious examples. Many examples are available in other books (e.g. [13]) and I am also a believer that users should simulate their own examples to learn effectively*. It seemed better to use the limited space available to say something which complements, rather than duplicates, the existing literature. There is not space to present every algorithm although one notable omission could be considered robust MPC algorithms based on other norms (such as the one and infinity norms). Again these are discussed in [13] but I do not believe have gained much acceptance as they are not as easily tuned as 2-norm based algorithms.

The hope behind this book is that readers will go away with good insight into MPC design and hence empowered to modify the algorithm to suit their own needs rather than being reliant on well known algorithms or the available commercial products. Moreover, if an implementation is performing poorly, they will be better placed to understand why.

> **Summary:** The beauty of MPC is its flexibility and yet relative simplicity of concept. Once fully appreciated, it is often straightforward to modify the algorithm for the specific requirements of each application.

---

*I intend to produce a simple suite of MATLAB programs which readers can use to form their own designs.

A

Appendix: Numerical examples and questions

This appendix gives some numerical data to help students test their code and/or understanding as well as some guidelines in setting questions for teaching staff. The intent is to make software available[*] so that users can develop and simulate their own controllers using some of the strategies in this book.

A.1 Numerical examples of prediction

This section gives some simple numerical examples of how to predict and form a nominal MPC law.

In general when writing your own code you need to devise your own consistency tests to ensure that the programmes are working correctly. For instance: (i) compare the predictions with your model to what arises from an open-loop simulation with arbitrary inputs; (ii) compare the actual closed-loop behaviour, as tuning varies with what you expect intuitively. Make sure you can explain any apparent anomalies.

A.1.1 State-space predictions of eqn.(3.10)

The predictions are given from

$$
\underbrace{\begin{bmatrix} \mathbf{x}_{k+1} \\ \mathbf{x}_{k+2} \\ \mathbf{x}_{k+3} \\ \vdots \\ \mathbf{x}_{k+n_y} \end{bmatrix}}_{\underset{\rightarrow k}{\mathbf{x}}} = \underbrace{\begin{bmatrix} A \\ A^2 \\ A^3 \\ \vdots \\ A^{n_y} \end{bmatrix}}_{P_{xx}} \mathbf{x}_k + \underbrace{\begin{bmatrix} B & 0 & 0 & \ldots \\ AB & B & 0 & \ldots \\ A^2B & AB & B & \ldots \\ \vdots & \vdots & \vdots & \vdots \\ A^{n_y-1}B & A^{n_y-2}B & A^{n_y-3}B & \ldots \end{bmatrix}}_{H_x} \underbrace{\begin{bmatrix} \mathbf{u}_k \\ \mathbf{u}_{k+1} \\ \mathbf{u}_{k+2} \\ \vdots \\ \mathbf{u}_{k+n_y-1} \end{bmatrix}}_{\underset{\rightarrow k-1}{\mathbf{u}}} \quad (A.1)
$$

---

[*]Contact the author or publisher if you wish to find out details about this.

and

$$
\underbrace{\begin{bmatrix} \mathbf{y}_{k+1} \\ \mathbf{y}_{k+2} \\ \mathbf{y}_{k+3} \\ \vdots \\ \mathbf{y}_{k+n_y} \end{bmatrix}}_{\underset{\to k}{\mathbf{y}}} = \underbrace{\begin{bmatrix} CA \\ CA^2 \\ CA^3 \\ \vdots \\ CA^{n_y} \end{bmatrix}}_{P} \mathbf{x}_k + \underbrace{\begin{bmatrix} CB & 0 & 0 & \dots \\ CAB & CB & 0 & \dots \\ CA^2B & CAB & CB & \dots \\ \vdots & \vdots & \vdots & \vdots \\ CA^{n_y-1}B & CA^{n_y-2}B & CA^{n_y-3}B & \dots \end{bmatrix}}_{H} \underset{\to k-1}{\mathbf{u}} \tag{A.2}
$$

Take a state-space model

$$
\mathbf{x}_{k+1} = A\mathbf{x}_k + B\mathbf{u}_k; \quad \mathbf{y}_{k+1} = C\mathbf{x}_{k+1} \tag{A.3}
$$

where

$$
A = \begin{bmatrix} 0.9 & -0.5 \\ 0 & 0.8 \end{bmatrix}; \quad B = \begin{bmatrix} 1 & 1 \\ -2 & 0 \end{bmatrix}; \quad C = \begin{bmatrix} 2 & 0.5 \\ -1 & 1 \end{bmatrix}
$$

Then one can define the matrices (for $n_y = 4$, $n_u = 2$) as follows:

$$
P_{xx} = \begin{bmatrix} 0.9 & -0.5 \\ 0 & 0.8 \\ 0.81 & -0.85 \\ 0 & 0.64 \\ 0.729 & -1.085 \\ 0 & 0.512 \\ 0.6561 & -1.2325 \\ 0 & 0.4096 \end{bmatrix}; \quad H_x = \begin{bmatrix} 1 & 1 & 0 & 0 \\ -2 & 0 & 0 & 0 \\ 1.9 & 0.9 & 1 & 1 \\ -1.6 & 0 & -2 & 0 \\ 2.51 & 0.81 & 1.9 & 0.9 \\ -1.28 & 0 & -1.6 & 0 \\ 2.899 & 0.729 & 2.51 & 0.81 \\ -1.024 & 0 & -1.28 & 0 \end{bmatrix}
$$

$$
P = \begin{bmatrix} 1.8 & -0.6 \\ -0.9 & 1.3 \\ 1.62 & -1.38 \\ -0.81 & 1.49 \\ 1.458 & -1.914 \\ -0.729 & 1.597 \\ 1.3122 & -2.2602 \\ -0.6561 & 1.6421 \end{bmatrix}; \quad H = \begin{bmatrix} 1 & 2 & 0 & 0 \\ -3 & -1 & & 00 \\ 3 & 1.8 & 1 & 2 \\ -3.5 & -0.9 & -3 & -1 \\ 4.38 & 1.62 & 3 & 1.8 \\ -3.79 & -0.81 & -3.5 & -0.9 \\ 5.286 & 1.458 & 4.38 & 1.62 \\ -3.923 & -0.729 & -3.79 & -0.81 \end{bmatrix}
$$

A.1.2   Prediction with transfer function models using eqn.(3.21)

Consider the difference equation model

$$
y_{k+1} = 1.3y_k - 0.4y_{k-1} + u_k - 2u_{k-1} \tag{A.4}
$$

First write this in incremental form of eqn.(3.14) as

$$y_{k+1} = 2.3y_k - 1.7y_{k-1} + 0.4y_{k-2} + \Delta u_k - 2\Delta u_{k-1} \tag{A.5}$$

Then substitute into expressions (3.21) using:

$$C_A = \begin{bmatrix} 1 & 0 & 0 & 0 \\ -2.3 & 1 & 0 & 0 \\ 1.7 & -2.3 & 1 & 0 \\ -0.4 & 1.7 & -2.3 & 1 \end{bmatrix}; \quad H_A = \begin{bmatrix} -2.3 & 1.7 & -0.4 \\ 1.7 & -0.4 & 0 \\ -0.4 & 0 & 0 \\ 0 & 0 & 0 \end{bmatrix}$$

$$C_{zb} = \begin{bmatrix} 1 & 0 & 0 & 0 \\ -2 & 1 & 0 & 0 \\ 0 & -2 & 1 & 0 \\ 0 & 0 & -2 & 1 \end{bmatrix}; \quad H_{zb} = \begin{bmatrix} -2 \\ 0 \\ 0 \\ 0 \end{bmatrix}$$

Hence (given $n_u = 2$)

$$H = \begin{bmatrix} 1 & 0 \\ 0.3 & 1 \\ -1.01 & 0.3 \\ -2.433 & -1.01 \end{bmatrix}; \quad P = \begin{bmatrix} -2 \\ -4.6 \\ -7.18 \\ -9.494 \end{bmatrix}; \quad Q = \begin{bmatrix} 2.3 & -1.7 & 0.4 \\ 3.59 & -3.51 & 0.92 \\ 4.747 & -5.183 & 1.436 \\ 5.7351 & -6.6339 & 1.8988 \end{bmatrix}$$

A.2 Numerical examples of control laws

In this section it is assumed for simplicity that the weighting on the controls is unity and that $n_y = 4$, $n_u = 2$. The reader is reminded that these horizons are small to facilitate numerical illustration but may not be good choices in general.

A.2.1 Control law of eqn. (4.53)

For convenience let the weighting matrix Q comprise blocks of the form $C^T C$ so that the optimal control law is equivalent to

$$K = [H^T H + I]^{-1} H^T P \tag{A.6}$$

Then substituting in the values from Section A.1.1 gives

$$K = \begin{bmatrix} 0.1045 & -0.3588 \\ 0.6726 & -0.1212 \end{bmatrix}$$

A.2.2   Control law of eqn. (4.25)

Take the prediction equations of Section A.1.2 (above and substitute into the control law of eqn.(4.25). Then

$$
\begin{aligned}
Pr &= [0.141 \ -0.069 \ -0.1758 \ -0.2307] \\
\check{D}_k &= 3.4872 \\
N_k &= [-2.0805 \ 2.4436 \ -0.6974]
\end{aligned}
\tag{A.7}
$$

A.3   Typical questions for tutors

Most questions on MPC either test insight, and hence are essay type questions, or require coding and simulation, and hence have no simple answers. As a consequence there are not many questions in this book. Most numerical questions are not pen and paper exercises, as predicting, even with SISO models, is numerically intensive for any horizon beyond 2 or 3. As such, there is a limit to the use of numerical questions which cannot be backed up by computer simulation. In order to compute prediction models and controllers for more complex cases the reader is advised to ask for the MATLAB software by contacting the publisher.

Summary: Easy questions for a tutor to set are:

- Descriptive type questions asking students to demonstrate understanding of the comments made in the summary boxes throughout the book.

- Coursework assignments where students could be asked to produce simulations demonstrating the validity, or limitations, of comments in the summary boxes.

- Numerical questions where students are asked to compute prediction equations, constraint equations, performance indices and/or control laws for low order examples with small horizons.

- MATLAB software will be provided to tutors so that they can produce prediction equations, constraint equations and do simulations and hence validate their answers. The software will come with some instructions and moreover will not require any MATLAB toolboxes other than the standard control toolbox (which comes with student MATLAB).

A.3.1 Example tutorial questions

Usually intelligent readers want to set their own questions and develop their own scenarios in order to test understanding, and therefore this book contains only a few illustrative problems. The tutor would need to supply the model data but could use the software provided to calculate any numerical answers. The material beyond Chapter 7 is largely beyond master's degree level and hence readers of these should focus mainly on the summaries and deriving their own problems to test understanding.

1. **Summary boxes:** Choose any summary box and base a question on it; for instance from Section 1.2.1.

   Discuss with illustrations why the use of prediction is beneficial in creating good control strategies.

   Or from Section 1.2.4:

   What guidelines should be used when selecting a prediction horizon?

2. **Prediction equations:** Define a model, transfer function, state-space or FIR, then specify whether the d.o.f. are the inputs or the input increments and finally pose questions like[†]:

   For the given model, find the prediction equations with $n_y = 3$ and $n_u = 2$.

   What is closed-loop prediction and when is its usage advisable?

3. **Control law and stability:** Define a prediction equation, for instance give the values of H, P as in eqn.(3.1) or any other prediction equation from Chapter 3. Then ask:

   For the performance index and predictions given, find the corresponding predictive control law and show how you would test stability in the nominal case.

4. **Tuning:** This is a difficult topic to set a numerical exam question on, as it would usually require significant computation. However, it could be ideal for coursework assignments.

   Using the software provided, illustrate and explain the effects of the tuning parameters n_y, n_u, λ on several examples.

5. **Sensitivity:** One could focus questions on what tools the designer has available to affect sensitivity. For instance:

   Derive the loop sensitivity with a GPC (or other) control law and discuss how the T-filter can be used to improve sensitivity.

[†]The software provided can be used to compute the numerical answer.

6. **Stability and performance guarantees:** Ask questions which test the students' ability of form a performance index with an infinite horizon and also to prove that infinite horizons allow stability guarantees. One could also ask more searching questions on tuning.

   Demonstrate why infinite output horizons facilitate stability and performance guarantees. How would the corresponding performance index be calculated?

   Describe dual mode control in the context of predictive control? Include detailed working on how a dual mode controller could be computed.

7. **Constraint handling:** Ask questions requiring students to construct the inequalities which ensure predicted constraint satisfaction. Ask for discussion on how this might impact on closed-loop stability or how the resulting optimisation might be solved. For instance:

   Given upper and lower limits on the input and input increments, define the constraint equation in the form $C\Delta\underset{\rightarrow}{\mathbf{u}} - \mathbf{d} \leq 0$ for $n_u = 3$.

A.3.2 Example exam questions

Q1 (a) You are given a system that has 2 inputs $\mathbf{u} = [u_1, u_2]^T$ which are subject to upper and lower limits of $[1,2]^T$ and $[-2,-3]^T$, respectively, and rate constraints $[0.5, 0.5]^T$. For a control horizon of $n_u = 3$, define the matrix C and vector $\mathbf{d}$ in the linear inequalities $C\Delta\underset{\rightarrow}{\mathbf{u}} - \mathbf{d} \leq 0$ which test for constraint satisfaction at sampling instant k.

   (b) You are given predictions that have the form

$$\underset{\rightarrow}{y}_k = H\Delta\underset{\rightarrow}{u}_{k-1} + P\Delta\underset{\leftarrow}{u}_{k-1} + Q\underset{\leftarrow}{y}_k \tag{A.8}$$

   where $\mathbf{y}$ is the system output. Hence define the quadratic programming problem whose solution gives the GPC control law for performance index:

$$J = \|\underset{\rightarrow}{\mathbf{r}} - \underset{\rightarrow}{\mathbf{y}}\|_2^2 + \lambda\|\Delta\underset{\rightarrow}{\mathbf{u}}\|_2^2 \tag{A.9}$$

   What is the unconstrained optimal control law?

   (c) In general a system may also have constraints on the outputs and other states. How might the objective be altered in the case that the constraints are not consistent.

Q2 The nominal process is given by a SISO CARIMA model

$$a(z)y_k = b(z)u_k + \frac{T(z)}{\Delta}\zeta_k \tag{A.10}$$

Given that $T(z) = 1$, the unbiased predictions take the form

$$\underset{\rightarrow k}{y} = H\Delta\underset{\rightarrow k-1}{u} + P\Delta\underset{\leftarrow k-1}{u} + Q\underset{\leftarrow k}{y} \qquad (\text{A.11})$$

and hence the predictive control law (SISO case) has the following form:

$$D_k(z)\Delta u_k = P_r(z)r_{k+1} - N_k(z)y_k \qquad (\text{A.12})$$

(a) Without giving all the working, state how the prediction structure varies with $T(z) \neq 1$.

(b) Hence, state how the control law structure differs from that in (A.12) when $T(z) \neq 1$.

(c) Plot a closed-loop diagram illustrating the constraint free implementation of GPC and hence contrast the transference from measurement noise to input for the control laws with and without a T-filter. Argue why the T-filter is popular and necessary in practice.

(d) A T-filter is a form of Youla parameterisation. Illustrate another means of incorporating a Youla parameter into the GPC control law and show how it can be used to improve sensitivity without affecting nominal tracking.

Q3 (a) Why do you think GPC is ideally suited as a control design technique for the multivariable case and the constrained case.

(b) A GPC control law for the model $D\mathbf{y} = N\mathbf{u}$ (MIMO case) takes the form

$$D_k\Delta\mathbf{u} = P_r\mathbf{r} - N_k\mathbf{y} \qquad (\text{A.13})$$

Show, with your working, how to compute the closed-loop poles.

(c) It is known that for an arbitrary choice of tuning parameters, GPC may not give good performance or even be stabilising. Suggest and justify some guidelines.

(d) Give an overview of how GPC can be modified to give a guarantee of stability.

Q4 (a) You are given a process $D\mathbf{y} = N\mathbf{u}$ and a GPC control law of the form

$$D_k\Delta\mathbf{u} = P_r\mathbf{r} - N_k\mathbf{y}$$

where $\mathbf{r}, \mathbf{y}, \mathbf{u}$ are the set point, output and input respectively.

Sketch the closed-loop diagram with output measurement noise and hence compute the sensitivity S of the input to output measurement noise.

The sensitivity is too high. Discuss what changes you could make to reduce this sensitivity and illustrate with the key equations.

(b) Prove that the use of infinite output horizons in the GPC performance index

$$J = \sum_{i=1}^{n_y} \|W_y(\mathbf{r}_{k+i} - \mathbf{y}_{k+i})\|_2^2 + \lambda \sum_{i=0}^{n_u-1} \|W_u(\Delta\mathbf{u}_{k+i})\|_2^2$$

guarantees closed-loop stability in the nominal case.

Given the limits of on-line computational load, suggest and justify some practical guidelines for choosing the tuning parameters.

Q5 (a) You are given a system that has 3 inputs $\mathbf{u} = [u_1, u_2, u_3]^T$. These are subject to upper and lower limits of $[3, 2, 2]^T$ and $[-2, -3, -1]^T$ respectively. There are no rate constraints. For a control horizon of $n_u = 3$, define the inequalities (in matrix form) which ensure constraint satisfaction of the input predictions.

(b) State constraints are also to be included. Define the prediction equations for the state for a prediction horizon $n_y = 3$ and the matrix inequalities ensuring constraint satisfaction for $n_y = 3$. You are given the following state-space model and upper limits:

$$\mathbf{x}_{k+1} = A\mathbf{x}_k + Bu_k; \quad \bar{\mathbf{x}} = \begin{bmatrix} 4 \\ 5 \\ 3 \end{bmatrix}$$

(c) Hence define the quadratic programming problem whose solution gives the GPC control law for performance index:

$$J = \sum_{i=1}^{3} \mathbf{x}_{k+1+i}^T Q\mathbf{x}_{k+1+i} + \mathbf{u}_{k+i}^T R\mathbf{u}_{k+i}$$

(d) Discuss why GPC is ideally suited as a control design technique for the multivariable case and the constrained case.

Q6 (a) First show how the robustness of a nominal predictive control law might be assessed. Second introduce and explain two methods which can be used to improve the loop sensitivity. Contrast the two methods briefly.

(b) Why can GPC with some certain tuning parameters give poor performance? Hence show how MPC can be tuned to give an a priori guarantee of the stability of the resulting closed-loop, even during constraint handling. Include some comment on potential weaknesses of these results.

B

Appendix: References

[1] J.C. Allwright, On mini-max model based predictive control, in workshop on Advances in model based predictive control, Oxford, 1993, pp246-255.

[2] J. Alvarez-Ramirez and R. Suarez, Global stabilisation of discrete time linear systems with bounded inputs, Int. J. ACSP, 10, 409-416, 1996.

[3] T.A. Badgwell, A robust model predictive control algorithm for stable linear plant, Proceedings American Control Conference, 1997.

[4] R.R. Bitmead, M. Gevers and V. Werrtz, Adaptive optimal control: The thinking man's GPC, Prentice Hall International, 1990.

[5] F. Blachini, Set invariance in control, Automatica, 35, 1747-1767, 1999.

[6] A. Bemporad, A. Casavola and E. Mosca, Nonlinear control of constrained linear systems via predictive reference management, Trans. IEEE AC, 42(3), 340-349, 1997.

[7] A. Bemporad, M. Morari, V. Dua and E.N. Pistokopoulos, The explicit linear quadractic regulator for constrained systems, Automatica, 38(1), 3-20, 2002.

[8] A. Bemporad, F. Borrelli and M. Morari, The explicit solution of constrained LP-based receding horizon control, Proceedings European Control Conference, Oporto, 2001.

[9] T.J.J. van den Boom and R.A. J de Vries, Cascade predictive controllers, IFAC World Congress, 1996.

[10] T.J.J. van den Boom and R.A. J de Vries, Constrained predictive control using a time varying youla parameter: a state space solution, European Control Conference 95, 1995.

[11] S. Boyd, L. El Ghaoui, E. Feron and V. Balakrishnan, Linear Matrix Inequalities, System and Control Theory, SIAM, Philadelphia, 1996.

[12] R. Boucher, Simulink model kindly supplied by Prismtech Ltd., Gateshead, UK, 1999.

[13] E.F. Camacho and C. Bordons, Model Predictive Control, Springer, 1999.

[14] M. Cannon and B. Kouvaritakis, Fast suboptimal predictive control with guaranteed stability, Systems and Control Letters, 35, 19-29, 1998.

[15] M. Cannon, B. Kouvaritakis and J.A. Rossiter, Efficient active set optimisation in triple mode MPC, IEEE Transactions on Automatic Control, 46(8), 1307-1313, 2001.

[16] A. Casavola, M. Giannelli and E. Mosca, Global predictive regulation of null-controllable input saturated linear systems, IEEE Trans AC, 44(11), 2226-2230, 1999.

[17] T. Chen and L. Qiu, H $_\infty$ design of general multirate sampled-data control systems, Automatica, 30(7), 1139-1152, 1994.

[18] L. Chisci and G. Zappa, Robust predictive control via controller invariant sets, IEEE Mediterranean Conference on Control and Automation, 1998.

[19] L. Chisci, J. A. Rossiter, G. Zappa Systems with persistent disturbances: Predictive control with restricted constraints, Automatica, 37(7), 2001.

[20] L. Chisci, J.A. Rossiter and G. Zappa, Robust Predictive Control with Restricted Constraints to cope with estimation errors, ADCHEM 2000 (Pisa), 2000.

[21] D.W. Clarke, C. Mohtadi and P.S. Tuffs, Generalised predictive control, Parts 1 and 2, Automatica, 23, 137-160, 1987.

[22] D.W. Clarke and R. Scattolini, Constrained receding horizon predictive control, Proceedings IEE, Pt. D, 138(4), 347-354, 1991.

[23] C.R. Cutler and B.L. Ramaker, Dynamic matrix control - a computer control algorithm, Proceedings American Control Conference, 1980.

[24] J.A. De Dona, M.M. Seron, D.Q. Mayne and G.C. Goodwin, Enlarged terminal sets guaranteeing stability of receding horizon control, Systems and Control Letters, 47, 57-63, 2002.

[25] B. De Moor, P. Van Overschee and W. Favoreel, Numerical algorithms for subspace state space system identification – an overview, Birkhauser, 1998.

[26] G. De Nicolao, L. Magni and R. Scattolini, Stabilising receding horizon control of non-linear time varing systems, IEEE Transactions on Automatic Control, 43, 1030-1036, 1998.

[27] W. Favoreel, B. De Moor and M. Gevers, SPC: subspace predictive control, Proceedings IFAC World Congress, Beijing, 235-240, 1999.

[28] A. Feuer and G.C. Goodwin, Sampling in Digital Processing and Control, Wiley, 1997.

[29] M. Fikar, S. Engell and P. Dostal. Design of infinite horizon predictive LQ controller, European Control Conference'97, Brussels, 1997.

[30] C.E. Garcia and M. Morari, Internal Model control 1. A unifying review and some new results, I&EC Process Design and Development, 21, 308-323, 1982.

[31] C.E. Garcia and A.M. Morshedi, Quadratic programming solution of dynamic matrix control (QDMC), Chem. Eng. Commun, 46, 73-87, 1986.

[32] C.E. Garcia, D.M. Prett and M. Morari, Model predictive control: theory and practice, a survey, Automatica, 25, 335-348, 1989.

[33] E.G. Gilbert and K. T. Tan, Linear systems with state and control constraints: the theory and application of maximal output admissable sets, IEEE Trans. AC, 36(9), 1008-1020, 1991.

[34] E.G. Gilbert and I. Kolmanovsky, Discrete-time reference governors and the non-linear control of systems with state and control constraint, Int. J. Rob. and Non-linear Control, 5, 487-504, 1995.

[35] E.G. Gilbert and I. Kolmanovsky, Discrete time reference governors for systems with state and control constraints and disturbance inputs, Proceedings Conference on Decision and Control, 1189-1194, 1995.

[36] E.G. Gilbert, I. Kolmanovsky and K.T.Tan, Nonlinear control of discrete time linear systems with state and control constraints: a reference governor with global convergence properties, Conference on Decision and Control, 144-149, 1994.

[37] E. G. Gilbert and I. Kolmanovsky, Discrete-time reference governors and the non-linear control of systems with state and control constraint, Int. J. Robust and Non-linear Control, 5, 487-504, 1995.

[38] E.G. Gilbert and I. Kolmanovsky, Fast reference governors for systems with state and control constraints and disturbance inputs, Int. Journal of Robust and Non-linear Control, 9, 1117-1141, 1999.

[39] C. Greco, G. Menga, E. Mosca and G. Zappa, Performance improvements of self-tuning controllers by multistep horizons: the MUSMAR approach, Automatica, 20, 681-699, 1984.

[40] J.R. Gossner, B. Kouvaritakis and J.A. Rosssiter, Stable Generalised predictive control in the presence of constraints and bounded disturbances, Automatica, 33(4), 551-568, 1997.

[41] J. R. Gossner, B. Kouvaritakis and J.A. Rossiter, Cautious stable predictive control: a guaranteed stable predictive control algorithm with low input activ-

ity and good robustness, 3rd IEEE Symposium on New Directions in Control and Automation, Cyprus, 243-250, Vol. II, 1995.

[42] J. R. Gossner, B. Kouvaritakis and J.A. Rossiter, Cautious stable predictive control: a guaranteed stable predictive control algorithm with low input activity and good robustness, International Journal of Control, 67(5), 675-697, 1997.

[43] M.J. Grimble, Polynomial systems approach to optimal linear filtering and prediction, International Journal of Control, 41, 1545-1566, 1985.

[44] M.J. Grimble, H_∞ optimal multichannel linear deconvolution filters, predictors and smoothers, International Journal of Control, 63(3), 519-533, 1986.

[45] P.-O. Gutman and M. Cwikel, An algorithm to find maximal state constraint sets for discrete time linear dynamical systems with bounded controls and states, IEEE Trans. AC, 32, 251-254, 1987.

[46] H. Hjalmarsson, M. Gevers and F. De Bruyne, For model based control design closed loop identification gives better performance, Automatica, 32, 1659-1673, 1996.

[47] H. Hjalmarsson, M. Gevers and S. Gunnarsson, Iterative feedback tuning: theory and applications, IEEE Control Systems Magazine, 16, 26-41, 1998.

[48] H. Hjalmarsson, Iterative feedback tuning: an overview, International journal of adaptive control and signal processing, 16, 373-395, 2002.

[49] Workshop on design and optimisation of restricted complexity controllers, Grenoble, January 2003.

[50] M. Hovd, J.H. Lee and M. Morari, Model requirements for model predictive control, European Control Conference, 2428-2433, 1991.

[51] K. Hrissagis and O.D. Crisalle, Mixed objective optimisation for robust predictive controller synthesis, Proceedings CPC-V, California, 1996.

[52] K. Hrissagis, O.D. Crisalle and M. Sznaier, 1995 Robust design of unconstrained predictive controllers, American Control Conference, 1995.

[53] P.P. Kanjilal, Adaptive prediction and predictive control, IEE Control Engineering Series 52, 1995.

[54] M. Kamrunnahar, D.G. Fisher and B. Huang, Model predictive control using an extended ARMarkov model, Journal of Process Control, 12, 123-129, 2002.

[55] S. Keerthi and E.G. Gilbert, Computation of minimum time feedback control laws for systems with state control constraints, IEEE Transactions on Automatic Control, 36, 1008-1020, 1988.

[56] E.C. Kerrigan and J.M. Maciejowski, Invariant sets for constrained nonlinear discrete-time systems with application to feasibility in model predictive control, Conference on Decision and Control, 2000.

[57] E.C. Kerrigan, Robust constraint satisfaction: Invariant sets and predictive control, Ph.D Thesis, Cambridge, UK, 2000.

[58] P.P. Khargonekar, K. Poolla and A. Tannenbaum, Robust control of linear time-invariant plants using periodic compensation, IEEE Transactions on Automatic Control, 30, 1088-1096, 1985.

[59] M.V. Kothare, V. Balakrishnan and M. Morari, Robust constrained model predictive control using linear matrix inequalities, Automatica, 32, 1361-1379, 1996.

[60] B. Kouvaritakis, J.A. Rossiter and A.O.T. Chang, Stable generalized predictive control: an algorithm with guaranteed stability, Proceedings IEE, Pt. D, 139(4), 349-262, 1992.

[61] B. Kouvaritakis, J.R. Gossner and J.A. Rossiter, Apriori stability condition for an arbitrary number of unstable poles, Automatica, 32(10), 1441-1446, 1996.

[62] B. Kouvaritakis, J.A. Rossiter and J.R.Gossner, Improved algorithm for multivariable stable GPC, Proceedings IEE Pt. D, 144(4), 309-312, 1997.

[63] B. Kouvaritakis and J.A. Rossiter, Multivariable stable generalized predictive control, Proc IEE Pt.D., 140(5), 364-372, 1993.

[64] B. Kouvaritakis, J.A. Rossiter and M. Cannon, Linear quadratic feasible predictive control, Automatica, 34(12), 1583-1592, 1998.

[65] B. Kouvaritakis, M. Cannon and J.A. Rossiter, Removing the need for QP in constrained predictive control, ADCHEM 2000 (Pisa), 2000.

[66] B. Kouvaritakis, J.A. Rossiter and J. Schuurmans, Efficient robust predictive control, IEEE Transactions on Automatic Control, 45(8), 1545-1549, 2000.

[67] B. Kouvaritakis, M.C. Cannon and J.A. Rossiter, Efficient active set optimisation in triple mode MPC, IEEE Transactions on Automatic Control, 46(8), 1307-1313, 2001.

[68] G.M. Kranc, Input output analysis of multi-rate feedback system, IRE Transactions on Automatic Control, 3, 21-28, 1957.

[69] W.H. Kwon and A.E. Pearson, A modified quadratic cost problem and feedback stabilisation of a linear system, IEEE Transactions on Automatic Control, 22(5), 838-842, 1978.

[70] I.D. Landau and A. Karimi, Recursive algorithms for identification in the closed loop: a unified approach and evaluation, Automatica, 33(8), 1499-1523, 1997.

[71] J.H. Lee, Recent advances in model predictive control and other related areas, CPC V (California), 1-21, 1996.

[72] J.H. Lee and M. Morari, Robust inferential control of multi-rate sampled data systems, Chem. Eng Sci., 47, 865-885, 1992.

[73] Y.I. Lee and B. Kouvaritakis, Linear matrix inequalities and polyhedral invariant sets in constrained robust predictive control, International Journal of Non-linear and Robust Control, 1998.

[74] D. Li, S.L. Shah and T. Chen, Identification of fast-rate models from multirate data, Interantional Journal of Control, 74(7), 680-689, 2001.

[75] D. Liu, S.L. Shah and D.G. Fisher, Multiple prediction models for long range predictive control, IFAC World Congress (Beijing), 1999.

[76] L. Ljung, System identification: theory for the user, Prentice Hall, 1987.

[77] A.G.J. MacFarlane and B. Kouvaritakis, A design technique for linear multivariable feedback systems, International Journal of Control, 25, 837-874, 1977.

[78] J.M. Maciejowski, Predictive control with constraints, Prentice Hall, 2001.

[79] J.M. Maciejowski, Multivariable Feedback Design, Addison-Wesley, 1989.

[80] J.M. Martin Sanchez and Rodellar, J., Adaptive predictive control; from concepts to plant optimisation, Prentice Hall International, 1996.

[81] D.Q. Mayne, Control of constrained dynamic systems, European Journal of Control, 7, 87-99, 2001.

[82] D.G. Meyer, A parametrization of stabilizing controllers for multirate sampled-data systems, IEEE Transactions on Automatic Control, 35, 233-236, 1990.

[83] H. Michalska and D. Mayne, Robust receding horizon control of constrained nonlinear systems, IEEE Transactions on Automatic Control, 38, 1623-1633, 1993.

[84] D.Q. Mayne, J.B. Rawlings, C.V. Rao and P.O.M. Scokaert, Constrained model predictive control: Stability and optimality, Automatica, 36, 789-814, 2000.

[85] J.A. Mendez, B. Kouvaritakis and J.A. Rossiter, State space approach to interpolation in MPC, International Journal of Robust and Nonlinear Control, 10, 27-38, 2000.

[86] M. Morari and J.H. Lee, Model predictive control: Past present and future. Computers and chemical engineering, 23(4), 667-682, 1999.

[87] E. Mosca and J. Zhang, Stable redesign of predictive control, Automatica, 28(6), 1229-1233, 1992.

[88] E. Mosca, Optimal Predictive and Adaptive Control, Prentice Hall, 1995.

[89] E. Mosca, G.Zappa and J. M. Lemos, Robustness of a multipredictor adaptive regulator: Musmar, Automatica, 25, 521-529, 1989.

[90] K.R. Muske and J. R. Rawlings, Model predictive control with linear models, AIChE Journal, 39(2), 262-287, 1993.

[91] J.P. Norton, An Introduction to Identification, Academic Press, 1992.

[92] T. Prez, G. C. Goodwin, M. M. Sern, Cheap fundamental limitation of input constrained linear systems, IFAC World Congress, 2002.

[93] C.V. Rao, S.J. Wright and J.B. Rawlings, Application of interior point methods to model preduictive control, Journal of Optimisation Theory and Applications, 99(3), 723-757, 1998.

[94] J.B. Rawlings and K.R. Muske, The stability of constrained receding horizon control, Trans. IEEE AC, 38, 1512-1516, 1993.

[95] J.B. Rawlings, E.S. Meadows and K.R. Muske, Nonlinear model predictive control: A tutorial and survey, Proceedings ADCHEM, 1994.

[96] J.B. Rawlings, Tutorial overview of model predictive control, IEEE Control Systems Magazine, 20(3), 38-52, 2000.

[97] J. Richalet, A. Rault, J.L. Testud and J. Papon, Model predictive heuristic control: applications to industrial processes, Automatica, 14(5), 413-428, 1978.

[98] J. Richalet, Pratique de la commande predictive, Hermes, 1993.

[99] J. Richalet, Plenary lecture at UKACC 2000, UK, 2000.

[100] O. J. Rojas and G. C. Goodwin, A simple antiwindup strategy for state constrained linear control, IFAC World Congress, 2002.

[101] J.A. Rossiter and B.Kouvaritakis, Constrained stable generalized predictive control, Proc IEE Pt.D, 140(4), 243-254 , 1993.

[102] J.A. Rossiter, Notes on multi-step ahead prediction based on the principle of concatenation, Proceedings IMechE., 207, 261-263, 1993.

[103] J.A. Rossiter, B.Kouvaritakis and J.R. Gossner, Feasibility and stability results for constrained stable generalised predictive control, 3rd IEEE Conference on Control Applications (Glasgow), 1885-1890, 1994.

[104] J.A. Rossiter and B. Kouvaritakis, Numerical robustness and efficiency of generalised predictive control algorithms with guaranted stability, Proceedings IEE Pt. D., 141(3), 154-162, 1994.

[105] J.A. Rossiter, B.Kouvaritakis and J.R. Gossner, Feasibility and stability results for constrained stable generalised predictive control, Automatica, 31(6), 863-877, 1995.

[106] J.A. Rossiter, B. Kouvaritakis and J.R. Gossner, Mixed objective constrained stable predictive control, Proceedings IEE, 142(4), 286-294, 1995.

[107] J.A. Rossiter, J.R. Gossner and B.Kouvaritakis, Infinite horizon stable predictive control, IEEE Transactions on Automatic Control, 41(10), 1522-1527, 1996.

[108] J.A. Rossiter, J.R. Gossner and B.Kouvaritakis, Gauranteeing feasibility in constrained stable generalised predictive control, Proceedings IEE Pt.D, 143(5), 463-469, 1996.

[109] J.A. Rossiter and B.G. Grinnell, Improving the tracking of GPC controllers, Proceedings IMechE., 210, 169-182, 1996.

[110] J.A. Rossiter, J.R. Gossner and B.Kouvaritakis, Constrained cautious stable predictive control, Proceedings IEE Pt.D, 144(4), 313-323, 1997.

[111] J.A. Rossiter, Predictive controllers with guaranteed stability and mean-level controllers for unstable plant, European Journal of Control, 3, 292-303, 1997.

[112] J.A. Rossiter, L. Chisci and A. Lombardi, Stabilizing predictive control algorithms in the presence of common factors, European Control Conference, 1997.

[113] J.A. Rossiter, M.J. Rice and B. Kouvaritakis, A numerically robust state-space approach to stable predictive control strategies, Automatica, 34, 65-73, 1998.

[114] J.A. Rossiter and B. Kouvaritakis, Youla parameter and robust predictive control with constraint handling, Workshop on Non-linear Predictive Control (Ascona), 1998.

[115] J.A. Rossiter, L.Chisci and B. Kouvaritakis, Optimal disturbance rejection via Youla-parameterisation in constrained LQ control, IEEE Mediterranean Conference on Control and Automation, 1998.

[116] J.A. Rossiter and L. Chisci, Disturbance rejection in constrained predictive control, Proceedings UKACC, 612-617, 1998.

[117] J.A. Rossiter and B. Kouvaritakis, Reducing computational load for LQ optimal predictive controllers, Proceedings UKACC, 606-611, 1998

[118] J.A. Rossiter, M.J.Rice, J. Schuurmanns and B. Kouvaritakis, A computationally efficient contrained predictive control law, American Control Conference, 1998.

[119] J. A. Rossiter and B. Kouvaritakis, Reference Governors and predictive control, American Control Conference, 1998.

[120] J.A. Rossiter, L. Yao and B. Kouvaritakis, Identification of prediction models for a non-linear power generation model, UKAmerican Control Conference Control Conference (Cambridge), 2000.

[121] J.A. Rossiter, L. Yao and B. Kouvaritakis, Application of MPC to a non-linear power generation model, Proceedings UKACC, 2000.

[122] J.A. Rossiter, B. Kouvaritakis and M. Cannon, Triple mode control in MPC, American Control Conference, 2000.

[123] J.A. Rossiter, B.Kouvaritakis and M.Cannon, Triple Mode MPC for enlarged stabilizable sets and improved performance, Conference on Decision and Control (Sydney), 2000.

[124] J.A. Rossiter, B. Kouvaritakis and L. Huaccho Huatuco, The benefits of implicit modelling for predictive control, Conference on Decision and Control, 2000.

[125] J. Schuurmanns and J.A. Rossiter, Robust piecewise linear control for polytopic systems with input constraints, IEE Proceedings Control Theory and Applications, 147(1), 13-18, 2000.

[126] J.A. Rossiter and B. Kouvaritakis, Modelling and implicit modelling for predictive control, International Journal of Control, 74(11), 1085-1095, 2001.

[127] J.A. Rossiter, Re-aligned models for prediction in MPC: a good thing or not?, Advanced Process Control 6, York, 63-70, 2001.

[128] J.A. Rossiter, B. Kouvaritakis and M. Cannon, Computationally efficient predictive control, ISSC, Maynooth, Ireland, 2001.

[129] J.A. Rossiter, B. Kouvaritakis and M. Cannon, Computationally efficient algorithms for constraint handling with guaranteed stability and near optimality, International Journal of Control, 74(17), 1678-1689, 2001.

[130] J.A. Rossiter and L. Chisci, An efficient quadratic programming algorithm for predictive control, European Control Conference, 2001.

[131] J.A. Rossiter, P.W. Neal and L. Yao, Applying predictive control to fossil fired power station, Proceedings Inst. MC, 24(3), 177-194, 2002.

[132] J.A. Rossiter and J. Richalet, Handling constraints with predictive functional control of unstable processes, American Control Conference, 2002.

[133] J.A. Rossiter and J. Richalet, Predictive functional control: alternative ways of prestabilising, IFAC World Congress, 2002.

[134] J.A. Rossiter, Improving the efficiency of multi-parametric quadratic programming, Workshop on Non-linear Predictive Control (Oxford), 2002.

[135] J.A. Rossiter, T.Chen and S.L.Shah, Developments in multi-rate predictive control, ADCHEM, (to appear), 2003.

[136] J.A. Rossiter, T. Chen and S.L. Shah, Improving the performance of dual rate control in the absence of a fast rate model, internal report, Dept. ACSE, University of Sheffield, 2003.

[137] P.O.M. Scokaert and J.B. Rawlings, Infinite horizon linear quadratic control with constraints, Proceedings IFAC World Congress, Vol. M, 109-114, 1996.

[138] P.O.M. Scokaert and J.B. Rawlings, Constrained linear quadratic regulation, IEEE Transactions on Automatic Control, 43(8), 1163-1168, 1998.

[139] J. Sheng, Generalized predictive control of multi-rate systems, Ph.D. thesis, University of Alberta, 2002.

[140] J. Sheng, T. Chen and S.L. Shah, Generalized predictive control for non-uniformly sampled systems, Journal of Process Control, 12, 875-885, 2002.

[141] D.S. Shook, C. Mohtadhi and S.L. Shah, Identification for long range predictive control, Proceedings IEE Pt-D, 138(1), 75-84, 1991.

[142] D.S. Shook, C. Mohtadhi and S.L. Shah, A control relevant identification strategy for GPC, IEEE Transactions on Automatic Control, 37(7), 975-980, 1992.

[143] S. Skogestad, K. Havre and T. Larsson, Control limitations for unstable plants, IFAC World Congress, 2002.

[144] R. Soeterboek, Predictive Control: a Unified Approach, Prentice Hall, 1992.

[145] M. Sznaier and M.J. Damborg, Heuristially enahnced feedback control of constrained discrete time linear systems, Automatica, 26(3), 521-532, 1990.

[146] K.T. Tan and E.G. Gilbert, Multimode controllers for linear discrete time systems with general state and control constraints, in Optimisation Techniques and Applications (World Scientific, Singapore), 433-442, 1992.

[147] A.K. Tangirala, D. Li, R.S. Patwardhan, S.L. Shah and T. Chen, Ripple free conditions for lifted multi-rate control systems, Automatica, 37, 1637-1645, 2001.

[148] J. Tse, J. Bentsman and N. Miller, Minimax long range parameter estimation, Conference on Decision and Control , 277-282, 1994.

[149] T.T.C. Tsang and D.W. Clarke, Generalised predictive control with input constraints, IEE Proceedings Pt. D, 6, 451-460, 1988.

[150] R.A. J de Vries and T.J.J. van den Boom, Robust stability constraints for predictive control, European Control Conference, 1997.

[151] R.A. J de Vries and T.J.J. van den Boom, Constrained robust predictive control, European Control Conference, 1995.

[152] B. Wams and T.J.J. van den Boom, Youla like parameterisations for predictive control of a class of linear time varying systems, American Control Conference, 1997.

[153] Z. Wan and M.V. Kothare, A computationally efficient formulation of robust predictive control using linear matrix inequalities, Proceedings of CPC-6, 2000.

[154] L. Wang, Use of orthonormal functions in continuous time MPC design, UKACC, 2000.

[155] A.G. Wills and W.P. Heath, Using a modified predictor corrector algorithm for model predictive control, IFAC World Congress, 2002.

[156] G.F. Wredenhagen and P. R. Belanger, Piecewise linear LQ control for systems with input constraints, Automatica, 30, 403-416, 1994.

[157] D.C. Youla and J.J. Bongiorno. A feedback theory of two-degree-of-freedom optimal Wiener-Hopf design, IEEE Transactions on Automatic Control, 30, 652-664, 1985.

[158] T.-W. Yoon and D.W. Clarke, Observer design in receding horizon predictive control, International Journal of Control, 61, 171-191, 1995.

[159] Z.Q. Zheng and M. Morari, Robust stability of constrained model predictive control, American Control Conference, 379-383, 1993.

Index

315